FLUID FLOW MEASUREMENT

FLUID FLOW MEASUREMENT

A Practical Guide to Accurate Flow Measurement

Third edition

PAUL J. LANASA AND E. LOY UPP[†]

AMSTERDAM • BOSTON • HEIDELBERG • LONDON • NEW YORK • OXFORD
PARIS • SAN DIEGO • SAN FRANCISCO • SINGAPORE • SYDNEY • TOKYO
Butterworth-Heinemann is an imprint of Elsevier

Butterworth-Heinemann is an imprint of Elsevier
225 Wyman Street, Waltham, MA 02451, USA
The Boulevard, Langford Lane, Kidlington, Oxford, OX5 1GB, UK

First edition 2002
Second edition 2002
Third edition 2014

Notices
No responsibility is assumed by the publisher for any injury and/or damage to persons
or property as a matter of products liability, negligence or otherwise, or from any use or
operation of any methods, products, instructions or ideas contained in the material herein.
Because of rapid advances in the medical sciences, in particular, independent verification
of diagnoses and drug dosages should be made

British Library Cataloguing-in-Publication Data
A catalogue record for this book is available from the British Library.

Library of Congress Cataloging-in-Publication Data
A catalog record for this book is available from the Library of Congress.

ISBN: 978-0-12-409524-3

For information on all Butterworth-Heinemann publications
visit our website at http://store.elsevier.com

Printed and bound in the United States of America

14 15 16 17 18   10 9 8 7 6 5 4 3 2 1

CONTENTS

DEDICATION

We dedicate this book to our families, particularly our wives, Carole LaNasa and Ann Upp, who assumed most of the responsibilities in raising our families while we worked and traveled in pursuit of our careers. And we express deepest appreciation to the companies—Tennessee Gas Pipeline, The Boeing Company, Daniel Industries (now the Daniel Division of Emerson Process Management), NuTech Industries, Ultra Field Measurement Company, The W.E.S.T. Corporation's CEESI Measurement Solutions, and CPL & Associates—whose assignments provided the opportunity for most of our flow measurement experience.

For over 48 years, we have helped solve flow measurement problems. During this time it has become apparent to us that good flow measurement is not a simple commodity to be selected solely by comparing product specifications. Rather, successful flow measurement results from application of good products with a full understanding of the equally important topics discussed in this book.

We subtitled the book "A practical guide to accurate flow measurement" and are quite confident that *practical* know-how comes only from a thorough understanding of fluid flow basics coupled with extensive experience. We have tried to share our experience and that of our peers through the examples and illustrations in the book. If our readers can make any contribution to reducing flow measurement uncertainties by application of the book's information, we will feel more than amply rewarded for the time and effort invested in writing it.

PREFACE

As noted in the preceding Dedication, the tendency to make flow measurement a highly theoretical and technical subject overlooks a basic tenet: *Practical application of meters, metering principles, and metering instrumentation and related equipment is the real key to quality measurement.* And that includes the regular maintenance by trained and experienced personnel with quality equipment required to keep flow measurement systems operating so as to achieve their full measurement potential.

We cannot begin to name the many friends who make up our background of experience. They include the pioneers in flow measurement, flow measurement design engineers, operating personnel—ranging from top management to the newest testers—academic and research based engineers and scientists, worldwide practitioners, theorists, and those just getting started in the business.

Our personal experience has been that understanding creates the most complete comprehension. Standing in front of a "class" as a "student" asks for an explanation of a point just covered, quickly and clearly separates what you have learned by rote from that which you truly understand. One finds out very rapidly what one really knows. Hopefully you will find that which you need to know and understand.

Why *another* book on flow measurement? Several factors motivated us. We have mentioned our emphasis on the *practical* side of the subject. Another reason is the large number of retirements and passings of experienced measurement personnel, including ourselves. And a third consideration is the tendency to make our various measurement standards "technically defensible"—but confusing.

We felt simply that a *practical guide* could be a useful project.

In the material covering standards, the brief overviews are coupled with our hope that interested readers will consult the documents and organizations listed for additional information. In the same vein, detailed theoretical discussions are left to such excellent sources as the latest edition of the *Flow Measurement Engineering Handbook* by R.W. Miller. Because of the extent of such detailed information, we present only outlines along with reference information for the reader's use.

We hope that enough practical information will be found in this book to help a reader analyze a flow problem to the extent that the other detailed references will become clear. We have tried to "demystify" flow measurement by breaking the subject into simple sections and discussing them in everyday terms. Each technology has its own terminology and jargon; that is why you will find many definitions and explanations of terms in the book.

In short, flow measurement is based in science, but successful application depends largely on the art of the practitioner. Too frequently we blindly follow the successful artist simply because "that is the way we've always done it." Industry experience the world over shows, however, that understanding *why* something is done can almost always generate better flow measurement.

Reference

Miller, R.W., Flow Measurement Engineering Handbook. 3rd Ed. McGraw-Hill, New York.

MEMORIAL

E. LOY UPP

Edmund Loy Upp, 79, of Sugar Land, TX, friend, colleague, and co-author passed away on June 24, 2007. E. Loy Upp was born September 10, 1927, in Stillwater, OK, to Faye and Dr. Charles Upp. He was the middle son of three sons raised in Baton Rouge, LA.

Mr. Upp graduated in 1944 from Louisiana State University with a Bachelor of Science Degree in Chemical Engineering. While in college, he was a member of the Sigma Chi Fraternity and served as Manager of Mike the Tiger. He was a lifetime LSU fan. After 50 years of dedication, he became a proud member and honoree of the LSU Gold Tiger Society.

Mr. Upp was well respected throughout his career in the oil and gas industry, and considered an expert in the field of flow measurement. He began his career at Tennessee Gas Pipeline Company, working as a measurement engineer. He traveled the world, offering his expertise in flow measurement to prominent universities and professional symposiums. Mr. Upp was the first President of the Gulf Coast Gas Measurement Society. In addition, he contributed articles/papers to the American Gas Association AGA-3, Gas Measurement Manual, ASME Fluid Meters, Sixth Edition, Gas Processors Association, Engineering Data Book and the American Petroleum Institute Petroleum Measurement Standards. He authored a prominent book on Fluid Flow Measurement.

Mr. Upp received many honors from the Petrochemical Industry, including: The Laurence Reid Award from the International School of Hydrocarbon Measurement, The AGA, The API Citation for Service, and the GPA recognition award. He concluded his 48 year gas and liquid measurement career as Vice President of Operations, Ultra Field Measurement Company. Mr. Upp previously served over 20 years as Director of Technology, Daniel Industries.

INTRODUCTION

Chapter Overview

The vast majority of this book relates to "conventional" flow-meters; for example, the admonition about single phase flow. Obviously, this comment does not apply to multiphase meters. Other exceptions are noted as they appear.

The book's general approach is to look first at basic principles, particularly with respect to differential and linear meters and the types used to measure fluid flow in the oil and gas industry. After a review of basic reference standards, "theory" is turned into "practice," followed by an overview of fluids and the fluid characteristics. "Flow" itself is examined next, followed by operating and maintenance concerns. Next, comments are offered on individual meters and associated equipment with a detailed review of the two classes of meters: differential and linear readout systems. Meter proving systems are covered in detail, followed by measurement data analysis and "lost and unaccounted for" procedures. The book concludes with a discussion of conversion to volumes, conversion of the volumes to billing numbers, and the audit procedures required to allow both parties to agree to the final measurement and money exchange.

Emphasis is not so much on individual meter details as on general measurement requirements and the types of meters available to solve particular problems.

Specifically, this first chapter presents some background information, overviews the requisites for "flow" and defines the major terms used throughout the book. Chapter 2 introduces various relevant subjects, starting with basic principles and fundamental equations. Chapter 3 details the types of fluid measurement: custody transfer and non-custody transfer. Chapter 4 is devoted entirely to listing the basic reference standards. Chapter 5 applies theory to the real world, and describes how various practical considerations make *effective* meter accuracy dependent on much more than simply the original manufacturer's specifications and meter calibration. Chapter 6 covers

Fluid Flow Measurement. ISBN: 978-0-12-409524-3

the limitations that fluid characteristics place on accurate flow measurements. Chapter 7 looks at flow in terms of the characteristics required, measurement units involved, and installation requirements for proper meter operation.

Chapter 8 reviews the necessary concerns in operating the meters properly, with examples of real problems found in the field. Chapter 9 covers the maintenance required for real metering systems to allow proper performance over time. Chapter 10 reviews meter characteristics, with comments on all the major meters used in the industry. Chapters 11 and 12 detail head and linear meters. Chapter 13 deals with related readout equipment. Chapter 14 discusses proving systems. Chapter 16 covers measurement data analysis and Chapter 15 covers material balance calculations and studies (i.e., lost and unaccounted for). Chapter 17 introduces the auditing required in oil and gas measurement.

Requisites of Flow Measurement

In this book, **fluids** are the common fluids (liquids, gases, steam, etc.) that are handled in the oil and gas industry, both in the pure state and in mixtures. However, each fluid of interest must be individually examined to determine whether it:

a. Is flashing or condensing;

b. Has well defined pressure, volume, temperature (PVT) relationships or density;

c. Has a predictable flow pattern based on Reynolds number;

d. Is Newtonian;

e. Contains any foreign material that will adversely affect the flow meter performance; (e.g., solids in liquids, liquids in gas);

f. Has a measurable analysis that changes slowly with time.

The flow should be examined to see if it:

a. Has a fairly constant rate or one that does not exceed the variation in flow allowed by the meter system response time;

b. Has a non-swirling pattern when entering the meter;

c. Is not two-phase or multiphase at the meter;

d. Is non-pulsating;

e. Is in a circular pipe running full;

f. Has provision for removing any trapped air (in liquid) or liquid (in gas) prior to entering the meter.

Certain meters may have special characteristics that can handle some of these problems, but they must be carefully evaluated to be sure of their usefulness for the fluid conditions actually encountered (Figure 1-1).

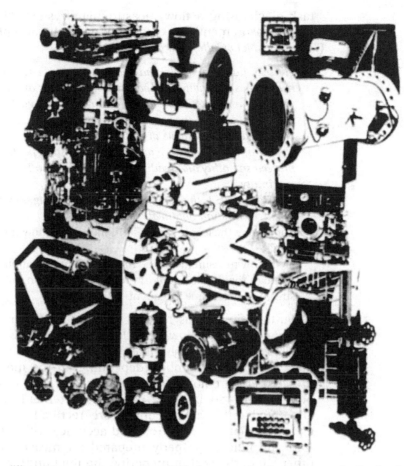

Figure 1-1 Many different types of meters are available for measuring flow. Proper selection involves a full understanding of all pertinent characteristics relative to a specific measurement job.

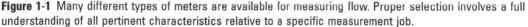

Measurement can usually be undertaken with any one of several meter systems, but certain meters have earned acceptance for specific applications based on their service record. This is an important factor in choosing a meter. Reference to industry standards and existing users within an industry are important points to review when choosing the best meter for a given application.

Background of Flow Measurement

The terms below form the background for fluid flow measurement, and should be understood before embarking on the task of choosing a flow measurement system. "Fluid," "flow" and "measurement" are defined in generally accepted terms (in *Webster's New Collegiate Dictionary*) as:

Fluid: 1. having particles that easily move and change their relative position without separation of the mass and that easily yield to pressure; 2. a substance (as a liquid or a gas) tending to flow or conform to the outline of its container.
Flow: 1. to issue or move in a stream; 2. to move with a continual change of place among the consistent particles; 3. to proceed smoothly and readily; 4. to have a smooth, uninterrupted continuity.
Measurement: 1. the act or process of measuring; 2. a figure, extent, or amount obtained by measuring.
Combining these into one definition for fluid flow measurement yields:
Fluid flow measurement: the measurement of smoothly moving particles that fill and conform to the piping in an uninterrupted stream to determine the quantity flowing.

Further limitations require that the fluids have a relatively steady state mass flow, are clean, homogeneous, Newtonian, and stable with a single phase non-swirling profile with some limit of Reynolds number (depending on the meter). If any of these criteria are not met, then the measurement tolerances can be affected, and in some cases measurement should not be attempted until the exceptions are rectified. These problems cannot be ignored, and expected accuracy will not be achieved until the fluid is properly prepared for measurement. On the other hand, the cost of preparing the fluid and/or the flow may sometimes outweigh the value of the flow measurement, and lower accuracy should be accepted.

History of Flow Measurement

Flow measurement has evolved over the years in response to demands to measure new products, measure old products under new conditions of flow, and for tightened accuracy requirements as the value of the fluid has increased (Figure 1-2).

Over 4,000 years ago, the Romans measured water flow from their aqueducts to each household to control allocation. The early Chinese measured the flow of salt water to brine pots used for producing the salt used as a seasoning. In each case, control over the process was the prime reason for the measurement.

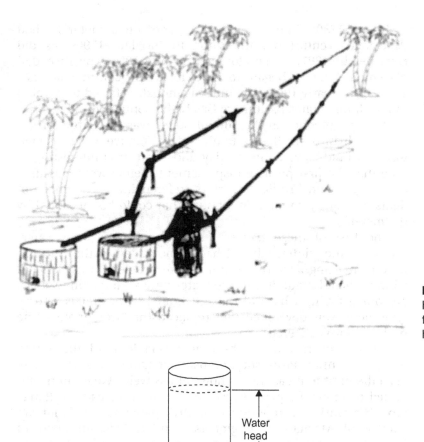

Figure 1-2 Flow measurement has probably existed in some form since man started handling fluids.

Water head

Figure 1-3 Bernoulli's theorem for orifice flow from a water pressure head was based on basic laws of physics relating velocity to distance and gravitational force.

Flow measurement for the purpose of determining billings for total flow developed later.

Well known names among the developers of the differential meter are Castelli and Tonicelli who, in the early 1600s, determined that the rate of flow was equal to the flow velocity times the area, and that discharge through an orifice varies with the square root of the head (pressure drop or differential).

In the early 1700s, Professor Poleni provided additional work on understanding the discharge of an orifice, and at about the same time Bernoulli developed the theorem upon which the hydraulic equations of head meters have been based ever since (Figure 1-3).

In the 1730s, Pitot published a paper on a meter he had developed. Venturi did the same in the late 1790s, as did Herschel in 1887. In London in the mid-1800s, positive displacement meters began to take form for commercial use. Then, in the early 1900s, the fuel-gas industry started to develop in the United States (Baltimore Gas Light Company).

An early practice in the United States was to charge for gas on a per-light basis; this certainly did not reduce waste, as customers would leave lights on day and night. It is interesting to note that the first positive displacement meters were classified as "5-light" and "10-light" meters, referencing the number of lights previously counted in a house that could be measured by the meter.

The first of these meters installed outdoors were water-sealed; in the winter, ethanol had to be added to the water to prevent freezing. One of the immediate problems was that not all the ethanol made it into the water baths—and some service personnel found it hard to make it home! In the 1800s a "dry" type meter was developed that replaced the "wet" meters (the prohibitionists cheered).

Rotary meters did not become available until the 1900s. About this time, Professor Robinson at Ohio State University used the pitot to measure gas flows at gas wells. Weymouth calibrated a series of square edged, thin plate orifices with flange taps. His work was reported in a 1912 paper to the American Society of Mechanical Engineers entitled "Measurement of Natural Gas." Similar tests were run on an orifice by Pugh and Cooper. Crude oil in this time period was measured by tank gauging. Tank gauging was the method used for storage tank batches from production to the final measurement of the refined products.

Around the same time, Professor Judd at Ohio State conducted tests on concentric, eccentric, and segmental orifice plates, and the forerunners of the present-day meter companies also conducted their own research, including Metric Metal Works (later American Meter), the Foxboro Company, and Pittsburgh Equitable (later Rockwell and Equimeter). To study the data and coordinate results, a committee of the American Gas Association (AGA) (1925) began additional testing. This work culminated in AGA Report No. 1, published in 1930 and reporting results to date for the test programs in progress. Work began immediately on AGA Report No. 2, which was published in 1935. The first AGA Report No. 3 was published in 1955.

Since that time, the quantity of additional data available is reflected in the reports published most recently. The AGA

Report No. 3, published in 1992, reflects a new discharge coefficient; a revision published in 2000 outlines new installation requirements. Current studies are evaluating the need for further revisions.

Paralleling these gas measurement efforts is the development of liquid meters for use in other areas of flow measurement, meters such as positive displacement, vortex shedding, ultrasonic, magnetic, turbine, and laser.

Flow measurement continues to change as the needs of the industry change. No end to such change and improvement is likely as long as mankind uses gas and liquid energy sources which require the measurement of flow.

Definition of Terms

Absolute Viscosity (mu) The absolute viscosity (mu) is the measure of a fluid's intermolecular cohesive force's resistance to shear per unit of time.

Accuracy The ability of a flow measuring system to indicate values closely, approximating the true value of the quantity measured.

Acoustical Twining The "organ pipe effect" (reaction of a piping length to a flow-pressure variation to alter the signal). Effects are evaluated based on acoustics.

Algorithm A step-by-step procedure for solving a problem, usually mathematical.

Ambient Conditions The conditions (pressure, temperature, humidity, etc.) externally surrounding a meter, instrument, transducer, etc.

Ambient Pressure/Temperature The pressure/temperature of the medium surrounding a flow meter and its transducing or recording equipment.

Analysis A test to define the components of the flowing fluid sample.

Base Conditions The conditions of temperature and pressure to which measured volumes are to be corrected (alternatively known as reference or standard conditions). The base conditions for the flow measurement of fluids, such as crude petroleum and its liquid products, having a vapor pressure equal to or less than atmospheric pressure at base temperature are:
In the United States:
Pressure: 14.696 psia (101.325 kPa)
Temperature: 60°F (15.56°C)
The International Standards Organization:
Pressure: 14.696 psia (101.325 kPa)
Temperature: 59°F (15°C)
For fluids, such as liquid hydrocarbons, which have a vapor pressure greater than atmospheric pressure at base temperature, the base pressure is customarily designated as the equilibrium vapor pressure at base temperature.
The base conditions for the flow measurement of natural gases are (in the USA):
Pressure: 14.73 psia (101.560 kPa)
Temperature: 60°F (15.56°C)
The International Standards Organization:
Pressure: 14.696 psia (101.325 kPa)
Temperature: 59°F (15°C)
For both liquid and gas applications, these base conditions can change from one country to the next, from one state to the next, or from one industry to the

next. Therefore, it is necessary that the base conditions be identified for "standard" volumetric flow measurement.

Beta Ratio The ratio of the measuring device diameter to the meter run diameter (i.e., orifice bore divided by inlet pipe bore).

Calibration of an Instrument or Meter The process or procedure of adjusting an instrument or a meter so that its indication or registration is in close agreement with a referenced standard.

Calorimeter An apparatus for measuring the heat content of a flowing fluid.

Certified Equipment Equipment with test and evaluations with a written certificate attesting to the device's accuracy.

Chart Auditing A visual review of field charts to find questionable dates.

Check Meter A meter added in series to check the billing meter.

Chilled Meter Test A test used to determine dew points (water and/or hydrocarbon) by passing the natural gas over a mirror while gradually reducing the temperature of the mirror until condensation forms.

Clock Rotation The time (in hours) needed to make a 360° chart rotation.

Coefficient of Discharge Empirically determined ratio from experimental data comparing measured and theoretical flow rates.

Compressibility The change in volume per unit of volume of a fluid caused by a change in pressure at constant temperature.

Condensing Reduction to a denser form of fluid (such as steam changing to water); a change in state from gas to liquid.

Condensing Point A point, measured in terms of pressure and temperature, at which condensation takes place.

Contaminants Undesirable materials in a flowing fluid that are defined by the quality requirements in a contract.

Control Signal (Flow) Information about flow rate that can be transmitted and used to control the flow.

Critical Flow Prover A test nozzle that is used to test the throughput of a gas meter where the linear velocity in the throat reaches the sonic velocity of the gas.

Critical Point That state at which the densities of the gas and liquid phases and all other properties become identical. This is an important correlating parameter for predicting fluid behavior.

Critical Pressure The pressure at which the critical point occurs.

Critical Temperature The temperature above which the fluid cannot exist as a liquid.

Custody Transfer A flow measurement whose purpose is to arrive at a volume for which payment is made/received as ownership is exchanged.

Dampening A procedure through which the magnitude of a fluctuating flow or pressure is reduced.

Density The density of a quantity of homogeneous fluid is the ratio of its mass to its volume. The density varies with temperature and pressure, and is therefore generally expressed as mass per unit volume at a specified temperature and pressure.

> **Density, Base** The mass per unit volume of the fluid being measured at base conditions (Tb, Pb).

> **Density, Relative (Gas)** The ratio of the specific weight of gas to the specific weight of air at the same conditions of pressure and temperature. (This term replaces the term "specific gravity" for a gas.)

> **Density, Relative (Liquid)** The ratio of a liquid's density at a given temperature to the density of pure water at a specific base temperature. (This term replaces the term "specific gravity" for a liquid.)

Diameter Ratio (Beta) The calculated orifice plate bore diameter (d) divided by the calculated meter tube internal diameter (D).

Differential Pressure The drop in pressure across a head device at specified pressure tap locations. It is normally measured in inches or millimeters of water.

Discharge Coefficients The ratio of the true flow to the theoretical flow. It corrects the theoretical equation for the influence of velocity profile, tap location, and the assumption of no energy loss with a flow area between 0.023 to 0.56 percent of the geometric area of the inlet pipe.

Electronic Flow Meter (EFM) An electronic flow meter readout system that calculates flow from transducers measuring the variables of the flow equation.

Element, Primary That part of a flow meter which is directly in contact with the flow stream.

Element, Secondary Indicating, recording, and transducing elements that measure related variables needed to calculate or correct the flow for variables in the flow equation.

Empirical Tests Tests based on data observed in experiments.

Energy The capacity for doing work.

    Energy, External Energy existing in the surroundings of a meter installation (normally heat or work energy).

    Energy, Flow Work Energy necessary to make the upstream pressure higher than that downstream, so that flow will occur.

    Energy, Heat Energy of the temperature of a substance.

    Energy, Internal Energy of a fluid due to its temperature and chemical makeup.

    Energy, Kinetic Energy of motion due to fluid velocity.

    Energy, Potential Energy due to the position or pressure of a fluid.

Equation of State The properties of a fluid are represented by equations that relate pressure, temperature, and volume. The usefulness of these equations depends on the database from which they were developed and the transport properties of the fluid to which they are applied.

Extension Tube (Pigtail) A piece of tubing placed on the end of a sample container used to move the point of pressure drop (point of cooling) away from the sample being acquired. See GPA 2166.

Flange Taps A pair of tap holes positioned as shown in Figure 1-4. The upstream tap center is located 1 inch (25.4 mm) upstream of the nearest plate face. The downstream tap center is located 1 inch (25.4 mm) downstream of the nearest plate face.

Flashing Liquids with a sudden increase in temperature and/or a drop in pressure vaporize to a gas flow at the point of change.

Floating Piston Cylinder A sample container that has a moving piston whose forces are balanced by a pre-charge pressure.

Flow

    Flow, Fluctuating A variation in flow rate that has a frequency that is lower than the meter-station frequency response.

    Flow, Ideal Flow that follows theoretical assumptions.

    Flow, Layered Flow that has sufficient liquid present to permit gas flow at a velocity above that for liquid flow at the bottom of a line. This flow is not accurately measured with currently available flow meters.

    Flow, Non-Fluctuating Flow that varies gradually in rate over long periods of time.

    Flow, Non-Swirling Flow with velocity components which move in straight lines with a swirl angle of less than 2° across the pipe.

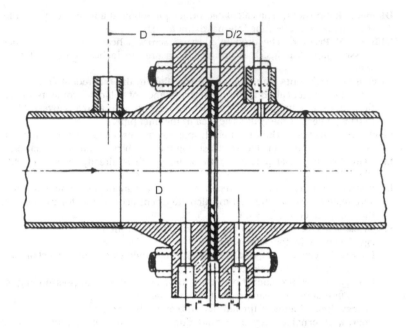

Figure 1-4 Location of flange taps.

Flow, Pulsating Variations in flow rate that have a frequency that is higher than the meter station frequency response.

Flow, Slug Flow that has sufficient liquid present to cause the liquid to collect in low spots and then "kick over" as a solid slug. This flow is not accurately measured by currently available flow meters.

Flow, Totalized The total flow over a stated period of time, such as per hour, per day, per month.

Flow Conditioning Preparing a flowing fluid so that it has no flow profile distortion or swirl.

Flow Nozzle A differential measuring device with a short cylinder that has a fluted approach section, as defined by the ASME standards.

Flow Profile A relationship of velocities in planes upstream of a meter that defines the condition of the flow into the meter.

Flow Proportional Composite Sampling The process of collecting gas over a period of time at a rate that is proportional to the pipeline flow rate.

Flow Rate The volume or mass of flow through a meter per unit time.

Flow Regime The characteristic flow behavior of a flow process.

Flow Temperature The average temperature of a flowing stream measured at a specified location in a metering system.

Fluid Flow Measurement The measurement of smoothly moving particles that fill and conform to the piping in an uninterrupted stream to determine the amount flowing.

Fluid Dynamics The mechanics of the flow forces and their relationship with the fluid motion and equilibrium.

Fluids, Dehydrated Fluids that normally have been separated into gas and liquid with the gas dried to the contract limit by a dehydration unit. (Normally the liquid is not dried, but it may be.)

Fluids, Separated Fluids that have been separated into gas and liquids at the temperature and pressure of the separating equipment.

Force Majeure Forces beyond the control of humans, normally from natural sources.

Frequency Response The ability of a measuring device to respond to the signal frequency applied to it within a specified limit.

Gas Laws These relate the volume, temperature and pressure of a gas; they are used to convert the volume at one pressure and temperature to that found under another set of conditions, such as flowing conditions to base conditions.

    Gas Law Boyle's Law states that the volume occupied by a given mass of gas varies inversely with the absolute pressure if the temperature remains constant.

    Gas Law Charles' Law states that the volume occupied by a given mass of gas varies directly with the absolute temperature if the pressure remains constant.

Gas Lift Injection of gas into a reservoir containing liquid to remove the liquid in the resulting production.

Gas Quality Refers to the physical characteristics determined by the composition (including non-hydrocarbon components, specific gravity, heating value, and dew points) of the natural gas.

Gas Sample Distortion Any effect that results in a sample that is not representative of the flowing gas stream.

Gas Sampling System The system intended to deliver a representative sample of natural gas from the pipeline to the analytical device.

Gaseous Phase The phase of a substance that occurs at or above the saturated vapor line of a phase diagram. It fills its container and has no level.

Gasoline Stripping Plant A separation plant designed to remove the heavier hydrocarbons from a gas stream.

Grade, Commercial Less-than-pure substance that must meet a composition limit. Although it is normally called by the name of its major component, it is actually a mixture.

Grade, Reagent Very pure substance that can be considered pure for calculation purposes.

Head Devices Meters that use the difference in elevation or pressure between two points in a fluid to calculate a flow rate.

Homogeneous Mix A mixture that is uniform throughout a flow stream mix, particularly important when sampling a flowing stream for analysis and the calculation of fluid characteristics.

Hydrates Ice-like compounds, formed by water and some hydrocarbons at temperatures that can be above freezing (32°F). They can collect and block a meter system's flow.

Hydrocarbon Dew Point The temperature at a specific pressure at which hydrocarbon vapor condensation begins.

Ideal Gas Law Relationship of pressure, temperature, and volume with no corrections for compressibility.

Integration To calculate the recorded lines on a chart for the period of chart rotation.

Internal Controls A company's rules of operation and the methods used to control these rules.

Lag Time In a sample system, the time required for a molecule to migrate from the inlet of the sample probe to the inlet of an analyzer.

Laminar Flow Flow at 2000 Reynolds number and lower; it has a parabolic profile.

Level Measurement Determination of a liquid level in a vessel.

Manometer A device that measures the height (head) of liquid in a tube at the point of measurement.

Mass The property of a body that measures the amount of material it contains and causes it to have weight in a gravitational field.

Mass Meter Meter that measures mass of a fluid based on a direct or indirect determination of the fluid's weight rate of flow.

Master Meter A meter whose accuracy has been determined, used in series with an operating meter to determine the operating meter's accuracy.

Material Balance A comparison of the amount of material measured into a process or pipeline compared with the amount of material measured out.

Measurement The act or process of determining the dimensions, capacity, or amount of something.

Meter, Dynamic Meters that continuously measure a flowing stream.

Meter Factor (MF) A number obtained by dividing the quantity of fluid measured by the primary mass flow system by the quantity indicated by the meter during calibration. For meters, it expresses the ratio of readout units to volume or mass units.

Meter Inspection This may be as simple as an external visual check, or be as complex as a complete internal inspection and calibration of the individual parts against standards and a throughput test.

Meter Proving The procedure required to determine the relationship between the "true" volume of fluid measured by prover and the volume indicated by the meter.

Meter, Static Meters that measure by batch from a flowing stream by fill and empty procedures.

Meter System All the elements needed to make up a flow meter, including the primary, secondary, and related measurements.

Meter Tube The upstream and downstream piping of a flow meter installation required to meet minimum requirements of diameter, length, configuration, and condition necessary to create a proper flow pattern through the meter.

Meter Tube Internal Diameter (D, Dm, Dr) The calculated internal diameter of a meter tube (D) is the inside diameter of the upstream section of the meter tube computed at the flow temperature (Tf); the calculated meter tube internal diameter (D) is used in the diameter ratio and Reynolds number equations. The measured meter tube internal diameter (Dm) is the inside diameter of the upstream section of the meter tube at the temperature of the meter tube at the time of internal diameter measurements determined as specified in API Chapter 14.3, Part 2. The reference meter tube internal diameter (Dr) is the inside diameter of the upstream section of the meter tube at the reference temperature (Tr) calculated as specified in API Chapter 14.3, Part 2. The reference meter tube internal diameter is the nominal, certified, or stamped meter tube diameter within the tolerance of API Chapter 14.3 Part 2, Section 5.1.3, and stated at the reference temperature Tr.

Mixture Laws A fluid's characteristics can be predicted from a knowledge of the individual components' characteristics. These mixture laws have limits of accuracy that must be evaluated before applying them.

Mobile Sampling System The system associated with a portable gas chromatograph.

Multiphase Flow Two or more phases (solid, liquid, gas, vapor) in the stream.

Newtonian Liquids Liquids that follow Newton's second law, which relates force, mass, length, and time. The flow meters covered in this book measure Newtonian fluids.

Non-pulsating (see Pulsation) Variations in flow and/or pressure that are below the frequency response of the meter.

Normal Condensation Caused by an increase in pressure or a decrease in temperature.

Normal Vaporization Caused by a decrease in pressure or an increase in temperature.

Nozzle A flow device with an inlet profile that is elliptical along its centerline and made to a specified standard; they are usually used for high-velocity flows. They are resistant to erosion because of their shape.

Orifice Plate A thin plate in which a circular concentric aperture (bore) has been machined. The orifice plate is described as a "thin plate" and "with a sharp edge," because the thickness of the plate material is small compared with the internal diameter of the measuring aperture (bore), and because the upstream edge of the measuring aperture is sharp and square.

Orifice Plate Bore Diameter (D, Dm, Dr) The calculated orifice plate bore diameter (D) is the internal diameter of the orifice plate measuring aperture (bore) computed at flowing temperature (Tf). The calculated orifice plate bore diameter (D) is used in the flow equation for the determination of flow rate. The measured orifice plate bore diameter (Dm) is the measured internal diameter of the orifice plate measuring aperture (bore) at the temperature of the orifice plate at the time of bore diameter measurements determined as specified in API Chapter 14.3, Part 2. The reference orifice plate bore diameter (Dr) is the internal diameter of the orifice plate measuring aperture at reference temperature (Tr), calculated as specified in API Chapter 14.3, Part 2. The reference orifice plate bore diameter is the nominal, certified, or stamped orifice plate bore diameter within the practical orifice plate bore diameter tolerance of API Chapter 14.3, Part 2, Table 2-1, and stated at the reference temperature Tr.

Orifice Plate Holder A pressure-containing piping element, such as a set of orifice flanges or orifice fitting, used to contain and position the orifice plate in the piping system.

Phase A state of matter such as solid, liquid, gas, or vapor.

Phase Change A change from one phase to another (such as liquid to gas). Most flow meters cannot measure at this condition.

Physical Constants The fundamental units adopted as primary measure values for time, mass (quantity of matter), distance, energy, and temperature.

Pipeline Quality Fluids that meet the quality requirements for contaminants as specified in the exchange contract; such as clean, non-corrosive, single phase, component limits, etc.

Pitot Probe An impact device with an inlet and return port that provides flow to a "hot loop" by converting velocity into a differential pressure.

Pressure The following terms pertain to different categories of pressure.

Pressure, Ambient The pressure of the surrounding atmosphere.

Pressure, Atmospheric The atmospheric pressure or pressure of one atmosphere. The normal atmosphere (atm) is 101.325 kPa (14.696 psia); the technical atmosphere (at) is 98,066.5 Pa (14.222 psia).

Pressure, Absolute The static pressure plus atmospheric pressure. (Note: calculations use absolute pressure values to determine flow.)

Pressure, Back, Turbine Meter The pressure measured at specified pipe diameters downstream from the turbine flow meter under operating conditions.

Pressure, Differential (dP) The static pressure difference measured between the upstream and the downstream flange taps.

Pressure, Gauge The pressure measured relative to atmospheric pressure (atmospheric pressure is taken as zero).

Pressure, Impact The pressure exerted by a moving fluid on a plane perpendicular to its direction of flow. It is measured along the flow axis.

Pressure Liquid, High-Vapor A liquid that, at the measurement or proving temperature of the meter, has a vapor pressure equal to or higher than atmospheric pressure (see low-vapor pressure liquid).

Pressure Liquid, Low-Vapor A liquid that, at the measurement or proving temperature of the meter, has a vapor pressure less than atmospheric pressure (see high-vapor pressure liquid).

Pressure Loss (Drop) The differential pressure in a flowing stream (which will vary with flow rate) between the inlet and outlet of a meter, flow straightener, valve, strainer, lengths of pipe, etc.

Pressure, Partial The pressure exerted by a single gaseous component of a mixture of gases.

Pressure, Static (Pf) Pressure in a fluid or system that is exerted normal to the surface on which it acts. In a moving fluid, the static pressure is measured at right angles to the direction of flow.

Pressure, Reid Vapor (RVP) The vapor pressure of a liquid at 100°F (37.78°C) as determined by ASTM D 323-58, Standard Method of Test for Vapor Pressure of Petroleum Products (Reid Method).

Pressure, Vapor (True) The term applied to the true pressure of a substance to distinguish it from partial pressure, gauge pressure, etc. The pressure measured relative to zero absolute pressure (vacuum).

Pressure, Velocity The component of the moving fluid pressure that is due to its velocity; commonly equal to the difference between the impact pressure and the static pressure (see pressure, impact and static).

Primary Element The primary element in orifice metering is defined as the orifice plate, orifice plate holder with its associated differential pressure sensing taps, and the meter tube.

Provers Devices of known volume used to prove a meter.

Proving Throughput Testing meter volume against a defined volume of a prover.

Pseudocritical A gas mixture's compressibility may be estimated by combining the characteristic critical pressures and temperatures of individual components based on their percentages and calculating an estimated critical condition for the mixture.

Pulsation A rapid, periodic, alternate increase and decrease of pressure and/or flow. The effect of this on a meter depends on the frequency of the pulsation and the frequency response of the meter.

Quality Requirements Limits of non-contract material contaminants in the fluid.

Real Gas Law Ideal gas law corrected for the effect of compressibility.

Recirculation Region ("Eddy") An area within a piping system out of the main flow where gas is not continually being replaced even though gas is flowing through the system.

Refined Products Products that have been processed from raw materials to remove impurities.

Representative Gas Sample Compositionally identical, or as near to identical as possible, to the gas sample source stream.

Residual Impurities Any substances, such as air or natural gas components, that are left in a sample cylinder.

Retrograde Condensation Caused by a decrease in pressure or increase in temperature.

Retrograde Vaporization Caused by an increase in pressure or decrease in temperature.

Reynolds number A dimensionless number defined as $(\rho\ d\ v)/\mu$ where ρ is density, d is the diameter of the pipe or device, v is the velocity of the fluids and μ is the viscosity—all in consistent units. Its usefulness is in correlating meter performance from one fluid to another.

Sample/Sampling

Sample Container Any container used to hold a natural gas sample. Typical sample containers are constant volume cylinders or floating piston cylinders.

Sample Loop The part of the sampling system that conveys the sample from the probe to the container or analytical device. It is typically external to the analysis device. This should not be confused with the sample loop that exists inside an analytical device such as a gas chromatograph.

Sample Probe A device which extends through the meter tube or piping, into the stream to be sampled.

Sample Source Refers to the stream being sampled.

Sampling A defined procedure for removing a part of the flowing stream that is intended to be representative of the total flowing stream composition.

Saturated Natural Gas Gas that will condense if the pressure is raised or the temperature is lowered. Water content saturated with water. Hydrocarbon content saturated with hydrocarbons.

Saturation A state of maximum concentration of a component in a fluid mixture at a given pressure and temperature.

Seal Pot A reservoir installed on each gauge line to maintain a constant leg on a pressure differential device or to isolate corrosive fluids from the differential device.

Secondary Equipment Equipment used to read the variables at the primary meter.

Shrinkage The amount of loss in apparent volume when two fluids are mixed; caused by the interaction of variably sized molecules.

Single Phase One phase (such as liquid without solids or gases present).

Single Phase Flow Natural gas flowing at a temperature above the hydrocarbon dew point and free of compressor oil, water, or other liquid or solid contaminants in the flow stream.

Slip Stream ("Hot Loop" or "Speed Loop") Provides a continuous flow of sample.

Slippage Fluid that leaks between the clearance of the meter rotors and the meter body.

Smart Transducers Transducers with the built-in ability to correct for variations in measured or ambient conditions; an important requirement for most flow meters and measuring devices.

Sour A fluid that contains corrosive compounds (often sulfur based).

Specific Gravity (see Density, Relative, Gas and Liquid)

Specific Weight The force (weight/unit area) with which a body under specified conditions is attracted by gravity.

Stacked Transducers The installation of two or more differential pressure transducers of different maximum ranges to measure differential pressure on an orifice meter intending to extend the flow range of the meter.

Standard The following terms pertain to categories of measurement standards.

Standard A measuring instrument intended to define, to represent physically or to reproduce the unit of measurement of a quantity (or a multiple or sub-multiple of that unit), in order to transmit it to other measuring instruments by comparison.

Standard, International A standard recognized by an international agreement to serve internationally as the basis for fixing the value of all other standards of the given quantity.

Standard, National A standard recognized by an official national decision as the basis for fixing the value, in a country, of all other standards of the given quantity. In general, the national standard in a country is also the primary standard.

Standard, Primary A standard of a particular measure that has the highest metrological qualities in a given field. Note: (1) The concept of a primary standard is equally valid for base units and for derived units. (2) The primary standard is never used directly for measurement other than for comparison with duplicate standards or reference standards.

Standard, Secondary A standard, the value of which is fixed by direct or indirect comparison with a primary standard or by means of a reference-value standard.

Standard, Working A standard which, when calibrated against a reference standard, is intended to verify working measuring instruments of lower accuracy.

Standards Organizations Industry/government committees that write standards (see the discussion of testing in Chapter 4).

Steam

Steam, Saturated The end point of the boiling process. It is the condition in which all liquid water has evaporated and the fluid is a gas. Being the end point of the boiling process, its pressure automatically defines its temperature, and conversely its temperature defines its pressure. Saturated steam is unstable; heat loss starts condensation; heat addition superheats; pressure loss superheats; pressure gain starts condensation.

Steam, Superheated Pressure decrease or heat added to saturated steam will produce superheated steam, which acts as a gas and follows general gas laws with increased sensitivity to temperature and pressure measurements.

Steam, Wet (Quality Steam) A two-phase fluid containing gaseous and liquid water. The quality number defines what proportion of the mixture is gaseous; for example, "95% quality steam" indicates that 95% by weight of the mixture is a gas; 5% is liquid water.

Sunburst Chart A recorded chart with a wide and variable differential recording with a pattern associated with a sun symbol.

Sweet Fluids containing no corrosive compounds.

Swirling Flow Flow in which the entire stream has a corkscrew motion as it passes through a pipeline or meter. Most flow meters require swirl to be removed before attempting measurement, although some ultrasonic and Coriolis type meters claim to handle some swirl without flow conditioning.

System Balances (see Material Balance) In a pipeline system this information is reflected in a "loss or unaccounted for" report.

Tank Gauging A defined procedure of measurement of fluids in tanks by level determination.

Tap Hole A hole drilled radially in the wall of a meter tube or orifice plate holder, the inside edge of which is flush and without any burrs.

Temperature Measurement (Tf) Flowing fluid temperature measured at the designated upstream or downstream location as specified in API MPMA Chapter 14.3, Part 2. In flow measurement applications in which the fluid velocity is well below sonic, it is common practice to inset a temperature-sensing device in the middle of the flowing stream to obtain the flowing temperature. For practical applications, the sensed temperature is assumed

Table 1-1 Comparison of Four Common Systems of Temperature Units

	Fahrenheit	Celsius	Rankine	Kelvin
Water boils	212°	100°	672°	373.15
Water freezes	32°	0°	452°	273.15
Difference between freeze and boil	180°	100°	180°	100
Absolute zero	−459.67°	−273.15°	0°	0

to be the static temperature of the flowing fluid. The use of flowing temperature in this part of the standard requires the temperature to be measured in degrees Fahrenheit, °F, or degrees Centigrade, °C. However, if the flowing temperature is used in an equation of state to determine the density of the flowing fluid, it may require that the °F or °C values be converted to absolute temperature values of degrees Rankine (°R), or kelvin (K) (Table 1-1).

Temperature Stratification At low flow rates, proper mixing does not take place; layers of flow have different temperatures, densities, and speed of sound. Proper mixing must be achieved to measure the fluid temperature.

Throughput Tests The passage at the flowing fluid through the operating meter compared to volume standard at the operating flow rate.

Transition Flow Flow, with a variable velocity profile, at Reynolds number between 2,000 and 4,000.

Turbulent Flow Flow above 4,000 Reynolds number with a relatively flat velocity profile.

Uncertainty A statistical statement of measurement accuracy based on statistically valid information that defines 95% of the data points (twice the standard deviation).

Vapor Phase This term, used interchangeably with "gas," has various shades of meaning. A vapor is normally a liquid at normal temperature and pressure, but becomes a gas at elevated temperatures. The term "vapor" is also sometimes used to indicate that liquid droplets may be present. In a strict technical sense, however, the terms are interchangeable.

Velocity Time rate of linear motion in a given direction.

Venturi A defined head metering device that has a tapered inlet and outlet with a constricted straight middle section.

Viscosity A fluid's property that measures the shearing stress that depends on flow velocity, density, area, and temperature—which in turn affects the flow pattern to a meter and hence measurement results.

Water Dew Point The temperature at a specific pressure at which water vapor condensation begins.

Weight The force with which a body is attracted by gravity.

Wetted Part The parts of a meter which are exposed to the flowing fluid.

2

BASIC FLOW MEASUREMENT LAWS

All of the following laws should be recognized and obeyed before flow measurement is attempted. Certain physical laws explain what happens in the "real" world. Some of these laws explain what happens when fluid flows in a pipeline, and these in turn explain what happens to a flowing stream as it goes through a meter. All variables in the equations must be in consistent units of measurement.

The law of "conservation of mass" states that the mass rate is constant. In other words, the amount of fluid moving through a meter is neither added to nor taken from as it progresses from point 1 to point 2 (Figure 2-1). This is also called the "Law of Continuity," and it can be written mathematically as follows:

$$M_1 = M_2 \qquad (2.1)$$

where:

M_1 = mass rate upstream;
M_2 = mass rate downstream.

Since mass rate equals fluid density multiplied by pipe area multiplied by fluid velocity, Equation 2.1 can be rewritten as:

$$\rho_1 A_1 V_1 = \rho_2 A_2 V_2 \qquad (2.2)$$

where:

ρ = fluid density at a designated point in the pipe;
A = pipe area at the designated point;
V = average velocity in the pipe at the designated point;
$1, 2$ = upstream and downstream positions.

In terms of volume rate this can be restated as:

$$Q = AV \qquad (2.3)$$

where:

Q = volume per unit time at flowing conditions;
A, V = as previously defined.

The law of "conservation of energy" states that all energy entering a system at point 1 is also in the system at point 2,

Fluid Flow Measurement. ISBN: 978-0-12-409524-3

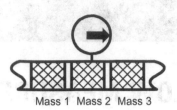

Mass 1 Mass 2 Mass 3

Figure 2-1 The amount of fluid flowing is constant at points 1 and 2.

even though one form of energy may be exchanged for another. (Note: the Bernoulli theorem relates the same physics in fluid mechanics.) The total energy in a system is made up of several types:

1. **Potential Energy** due to the fluid position or pressure.
2. **Flow Work Energy** required for the fluid to flow. The fluid immediately preceding that between point 1 and point 2 must be at a slightly higher pressure in order to exert a force on the volume between 1 and 2, so that this will flow.
3. **Kinetic Energy** (energy of motion) due to fluid velocity.
4. **Internal Energy** due to fluid temperature and chemical composition.
5. **External Energy** is energy exchanged with the fluid existing between point 1 and point 2 and its surroundings. These are normally heat and work energies.

The **Fluid Friction Law** states that energy is required to overcome friction and move fluid from point 1 to point 2. For the purpose of calculating flows, certain assumptions are made about the stability of the system energy under steady flow. The main energy concerns are the potential and kinetic energies (definitions 1 and 3); the others are either of no importance, do not change between position 1 and position 2, do not occur, or are taken care of by calibration procedures. A generalized statement of this energy balance is given below:

$$KE_1 + PE_1 = KE_2 + PE_2 \qquad (2.4)$$

Kinetic energy (KE) is energy of motion (velocity). Potential energy (PE) is energy of position (pressure).

In simple terms, this equation can be rewritten:

$$PE_1 + VE_1 = PE_2 + VE_2 \qquad (2.5)$$

where:
  PE = pressure energy;
  VE = velocity energy.

Equation 2.5 is the "ideal flow equation" for a restriction in a pipe. In real applications, however, certain corrections are necessary. The major equation correction is an efficiency factor called the "coefficient of discharge." This factor takes into account the difference between the ideal and the real world. The ideal equation states that 100% of the flow will pass an orifice with a given differential, when in fact empirical tests indicate that a lower fraction of the flow actually passes for a given differential—for example, about 60% with differential between flange taps on an orifice meter, 95% across a nozzle, and 98% across a Venturi (Figure 2-2). This is caused by the device's inefficiency or the loss from inefficiency caused by turbulence at the device where energy of pressure is not all converted to energy of motion. This factor has been determined by industry studies over the years and is reported as a "discharge coefficient."

Equations 2.4 and 2.5 assume no energy, such as heat, is added or removed from the stream between upstream and the meter itself. This is normally of small concern unless there is significant difference between the flowing and ambient temperatures (i.e., steam measurement), or when measuring a fluid whose volume is sensitive to very small temperature changes, such as near its critical temperature. (Three common examples are ethylene, carbon dioxide gas, and hot water near its boiling point.)

It is also assumed that no temperature change is caused by fluid expansion (because of the lower pressure in the meter) from the upstream pressure to the meter. The small pressure difference between the two locations normally makes this

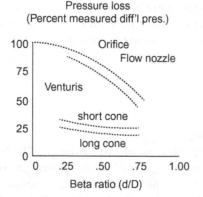

Pressure loss
(Percent measured diff'l pres.)

Figure 2-2 Orifice, flow nozzle, and Venturi meters all have permanent pressure losses somewhat less than 100% of full measurement differential; an ultrasonic meter is like an open section of pipe (non-intrusive) with no additional permanent pressure drop.

theoretical consideration insignificant. If there is a change in state (i.e., from liquid to gas or gas to liquid) then this "insignificant" temperature change is no longer insignificant. Furthermore, the volume occupied (assuming no mass holdup) is much greater in the gaseous phase than the liquid phase; volume ratios of gas to liquids are as much as several hundred for some common fluids. Because of these problems, flow measurement of flashing liquids or condensing gases should not be attempted.

Reynolds Number

The Reynolds number is a useful tool for relating how a meter will react to a variation in fluids from gases to liquids. Since an impossible amount of research would be required to test every meter on every fluid we wish to measure, it is desirable that a relationship between fluid factors be known. Reynolds' work in 1883 defines these relationships through his Reynolds number, which is defined by the equation:

$$Re = \frac{\rho D \upsilon}{\mu} \qquad (2.6)$$

where:
Re = Reynolds number, a dimensionless number;
ρ = density of the fluid;
D = diameter of the passage way;
v = velocity of the fluid;
μ = viscosity of the fluid.
Note: All parameters are given in the same units, so that when multiplied together they all cancel out, and the Reynolds number has no units. Units in the pound, foot, second system are shown below:
Re = no units;
ρ = #/cubic feet;
D = feet;
v = feet/sec;
μ = #/foot-sec.
Based on Reynolds' work, the flow profile (which affects all velocity-sensitive meters and some linear meters) has several important values. At values of 2,000 and below, the flow profile is bullet-shaped (parabolic). Between 2,000 and 4,000 the flow is in the transition region. At 4,000 and above the flow is in the turbulent flow area and the profiles are fairly flat. Thus, calculation of the Reynolds number will define the flow velocity

pattern and approximate limits of the meter's application. To completely define the meter's application there must be no deformed profiles, such as after an elbow or where upstream piping has imparted swirl to the stream.

These effects will be further discussed in the sections covering the description and application of different meters, in Chapters 8, 9, and 10, and the equations will be covered more thoroughly later in this book.

These equations can be combined and rewritten in simplified forms. However, it is important to recognize the assumptions which have been made, so that if a metering situation deviates from what has been assumed, a "flag will go up" to indicate that the effect of Reynolds number must be evaluated and treated.

Gas Laws

Gas properties are almost always measured at conditions other than standard or base conditions, so the measured values must be converted by calculation, using the gas laws, to those at the desired base conditions, to make the values comparable.

Boyle's Law states that the volume of a gas is inversely proportional to its pressure for an ideal gas at constant temperature (Figure 2-3).

$$V = Constant/P \tag{2.7}$$

where:

V = volume;

P = pressure.

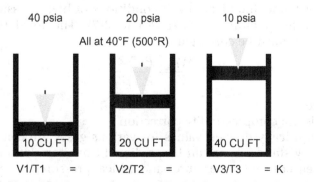

Figure 2-3 Diagrammatic representation of Boyle's Law, showing that volume is inversely proportional to pressure.

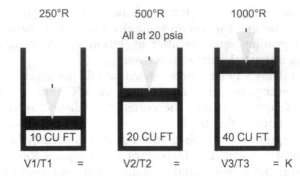

Figure 2-4 Diagrammatic representation of Charles' Law, showing that volume is proportional to temperature.

Charles' Law states that the volume of a gas is directly proportional to its temperature for an ideal gas at constant pressure (Figure 2-4).

$$V = Constant/T \qquad (2.8)$$

The Ideal Gas Law combines Boyle's and Charles' laws; it can be written:

$$\frac{V_b}{V_f} = \frac{P_f T_b}{P_b T_f} \qquad (2.9)$$

where:
$_b$ = base conditions;
$_f$ = flowing conditions.

The Real Gas Law (non-ideal) corrects for the fact that gases do not follow the ideal law at conditions of high pressure and/or low temperature (Figures 2-5 to 2-7). The Ideal Gas Law equation must be corrected to:

$$\frac{V_b}{V_f} = \frac{P_f T_b Z_b}{P_b T_f Z_f} \qquad (2.10)$$

where:
Z is the compressibility correction.

Empirically derived values for various gases are available in industry standards or can be predicted from correlations based on their critical temperatures and critical pressures.

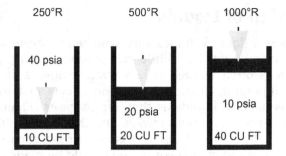

Figure 2-5 The Ideal Gas Law combines Boyle's and Charles' laws.

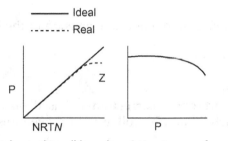

Figure 2-6 Ideal and actual conditions depart at extremes of pressure and temperature.

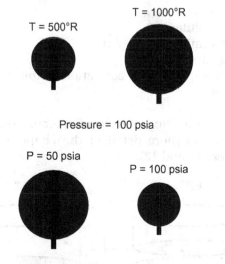

Figure 2-7 Gas-filled balloons also illustrate pressure/temperature/volume relationships.

Expansion of Liquids

Like gases, liquid volumes vary with temperature and pressure. Because liquids have little compressibility with pressure, this effect is often ignored unless temperatures approach the critical temperature (within 20%). The effects of temperature are not as large in liquids as they are in gases (Figure 2-8). If there is a large difference between the flowing and base temperatures or flowing and base pressures, or if a high degree of accuracy is desired, then a correction should be made.

The **temperature-effect correction** is based on cubical expansion for each liquid as follows:

$$V_b = V_f[1 + B(T_f - T_b)] \tag{2.11}$$

where:

B = the coefficient of cubical expansion of the liquid;
V = volume;
T = temperature;
$_b$ = base;
$_f$ = flowing.

The **pressure-effect correction** is based on compressibility effects for each liquid (this follows API procedures):

$$\left(\frac{1}{1 - [P - (P_e - P_b)]F} \right) \tag{2.12}$$

where:

P = pressure;
$_b$ = base, absolute;
$_e$ = equilibrium vapor, absolute;
$_f$ = flowing, absolute;
F = liquid compressibility correction factor.

If $P_b = P_e$, then:

$V_b = V_f$.

The specific equations for head meters and linear meters will be covered in more detail in the chapters which discuss them (Chapters 11 and 12).

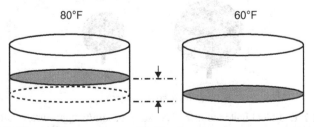

Figure 2-8 Effect of temperature on liquid volume.

Fundamental Flow Equation

The fundamental flow equation for differential devices is:

$$q_m = \frac{CE_v Y \pi d^2}{4} \sqrt{2g_c \rho_f \Delta P} \qquad (2.13)$$

where:

q = mass flow rate;
C = discharge coefficient of flow at the Reynolds number for the device in question;
Y = expansion factor (usually assumed to be 1.0 for liquids);
π = pi;
d = restriction diameter at flowing conditions;
g_c = dimensional conversion constant;
ρ_f = density of flowing fluid;
ΔP = difference in pressure during flow caused by the restriction;
E_v = velocity of approach factor.

(Note: Discharge coefficient values are available for specific restrictions with specific sets of pressure taps.)

Each standard and each differential device has equations that look different (notations are not the same), but they follow a basic relationship. The fundamental orifice meter mass flow equations in the various standards are:

API 14.3, Sec 3, Part 1

$$q_m = C_d E_v Y d^2 (\pi/4) \sqrt{2g_c \rho_{tp} \Delta P} \qquad (2.14)$$

AGA-3 1985

$$q_m = K Y_1 d^2 (\pi/4) \sqrt{2g_c \rho_{tp} \Delta P} \qquad (2.15)$$

ISO 5167

$$q_m = C E \varepsilon d^2 (\pi/4) \sqrt{2 \Delta P \rho} \qquad (2.16)$$

ASME

$$q_m = (\pi/4) C \varepsilon^2 \sqrt{\frac{2 \Delta P \rho}{1 - \beta^4}} \qquad (2.17)$$

In these equations:

$$C = C_d$$

$$E_v = E = [1/(1-\beta^4)]^{0.5}$$

$$Y = \varepsilon$$

$$2\Delta P = 2g_c \, \Delta p$$

$$K = C_d E_v = CE = C/[1/(1-\beta^4)]^{0.5}$$

The equation may be simplified to involve more easily measured variables (such as differential in inches of water rather than pounds per square inch), and some constants (such as π^4 and $2g_c$) may be reduced to numbers. If mixed units are used, corrections for these may also be included in the number. Details of this equation for each unit will be covered later in the appropriate section.

The "volume flow rate" may be derived from the mass flow:

$$q_b = \frac{q_m}{\rho_b} \tag{2.18}$$

where:

q_b = volume flow rate at base conditions;
q_m = mass flow rate;
ρ_b = density of fluid at base conditions.

Equations for non-head type volume meters are simpler since they basically reduce the volume determined by the meter under flowing conditions to that under base conditions, as was discussed in the section above on basic laws.

$$q_f F_b = q_b \tag{2.19}$$

where:

q_f = volume at flowing conditions;
F_b = factor to reduce from flowing to base (corrects for effects of compressibility, pressure, and temperature);
q_b = volume at base conditions.

If the meter measures mass, then there is no reduction to base required (a pound is a pound the world around!) and:

$$q_{mf} = q_{mb} \tag{2.20}$$

where:

q_{mf} = mass at flowing conditions;
q_{mb} = mass at base conditions.

References

American Gas Association, Gas Committee Report No. 3. 1515 Wilson Boulevard, Arlington, VA, 22209.

American Petroleum Institute. Chapter 14 of Manual of Petroleum Measurement Standards. 1220 L St., NW, Washington, DC 20005−4070.

American Society of Mechanical Engineers. Two Park Avenue, New York, NY 10016−5990.

International Organization for Standardization, ISO Central Secretariat, 1, ch. de la Voie-Creuse, CP 56, CH-1211 Geneva 20, Switzerland.

TYPES OF FLUID FLOW MEASUREMENT

Fluid flow measurement is divided into several types, since each type requires specific consideration of such factors as accuracy requirements, cost considerations, and use of the flow information to obtain the required end results.

What type of flow meter is best? A common question asked at flow measurement schools and seminars is: "What type of meter is best for my application?"

The answer obviously depends on many factors, but the first consideration, which is often ignored, should be about the fundamental nature of the fluid (liquid, gas, or steam) to be measured. Is there flashing or condensing? Are there well-defined pressure, volume, and temperature (PVT) relationships? Does a predictable flow pattern exist based on the Reynolds number (R_e)? Is the flow Newtonian? Is the fluid free of foreign materials that will affect meter performance (solids or gas in liquids, or liquids or solids in gas)? Does the fluid have a constant composition or slowly changing measurable analysis?

Flow characteristics are also important. Is there a fairly constant rate, or do rate variations fall within the meter's response time and measurement range? How about a laminar, irregular, swirling flow pattern? Single or multiphase at the meter inlet? Smooth or pulsating? Will the liquid flow fill the pipe?

Certain types of meters may have special characteristics to handle some of these problems, but extra care should be exercised in evaluating such equipment to ensure successful measurement. It is worth noting that most gases and many liquids that are routinely handled in the field are not as clean as often assumed. Time and again, measurement and meter evaluation projects have been seriously flawed by debris and crud buildup on metering components. An astute measurement specialist never assumes that the line is clean.

It is also important to consider the fluid's critical temperature and critical pressure. A meter's specified accuracy is invalid

Fluid Flow Measurement. ISBN: 978-0-12-409524-3

if the fluid to be measured exhibits large volume changes over minor temperature and pressure changes, which is the case near critical conditions.

At a meter station that measures product worth $1 million a day, an inaccuracy of ±0.2% represents $2,000 a day, or $730,000 a year—an amount that justifies considerable investment to improve flow measurement. The same error for a station measuring $1,000 worth of product a day represents only $2 a day, and the law of diminishing returns limits the investment that is justifiable to improve the measurement accuracy.

Custody Transfer

When money is to be exchanged the best flow measurement becomes important, so that the two parties to the transaction are treated fairly. The desired accuracy limit for flow measurement is 100% correct. However, no measurement is absolutely accurate; it is simply accurate to some limit.

In custody transfer metering, the constant awareness that flow measurement equates to dollars changes the perspective accordingly. The goal sought is that custody transfer measurement be converted to dollars ± zero.

Quantities for custody transfer are treated as absolute when they are billed. The responsibility for this measurement, then, is to reduce all inaccuracies to a minimum so that a measured quantity can be agreed upon for exchanging custody.

The accuracy of set point control measurement may be accepted at ±2%; operational measurement may require no more than ±5%, as contrasted with the ±0% target for custody transfer metering (Figure 3-1).

While ±0% is an ideal financial custody transfer goal, it is rarely achievable from a real or physical standpoint. Therefore, what measurement practitioners strive to achieve in custody transfer measurement is an equal sharing of risk across the measurement with as little uncertainty as is practicable.

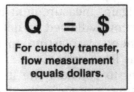

Q = $
For custody transfer,
flow measurement
equals dollars.

Figure 3-1 The flow meter is a cash register in custody transfer metering.

Measurement Contract Requirements

As stated, measurement becomes more complex when two parties must agree to the quantity of product exchanged and agree to pay/receive money based upon this quantity. To protect each party's interest, a contract is normally written that specifies all requirements for measurement of the delivered material, such as:

- Definitions used in the contract;
- Quantity of material;
- Point of delivery;
- Material properties;
- Measurement station design;
- Measurements to be made;
- Material quality;
- Price;
- Billing;
- Force majeure;
- Default or termination;
- Term;
- Warranty of title;
- Government requirements;
- Arbitration;
- Miscellaneous.

All of these items of interest should be settled prior to commencing measurement for custody transfer purposes. A number of these items are typical in any contract, but it is of value to expand on the ones that impact the measurement equipment and procedures.

Unfortunately, although intended to control the quality of the custody transfer exchange, contracts rarely contain sufficient measurement controls to achieve that objective given that the quality of the custody transfer measurement is a function of the following:

1. Selecting the appropriate metering device to achieve a given level of accuracy.
2. Installing the selected device in such a manner so that it can achieve its potential.
3. Operating the selected device in such a manner that it is capable of achieving its potential.
4. Processing the output(s) of the selected device in such a manner that it is capable of achieving its potential.
5. Maintaining the selected device in such a manner that it is capable of achieving its potential.

Contracts rarely address all of the above control requirements sufficiently, as will be seen in the following sections.

Quantity of Material

This will specify not only the quantity of the material to be measured by the seller, but also any rights the seller may have to quantities of material above or below the agreed-upon amount. This requires that the responsible measurement personnel are aware of these values, see that contract limits are being met, and ensure that the seller has the capability of meeting them.

Point of Delivery

The contract sets out the point of custody transfer. If the measurement point and the point of delivery are not the same, an agreement must be reached between the buyer and the seller for responsibilities for the material between these two points.

Material Properties

Limits are specified for certain basic properties, such as composition limits, pressure, and temperature, and the actions to be taken if the material is outside the limits.

Measurement Station Design

The ownership and responsibility of both the buyer and seller for the design, installation, maintenance, and operation of the meter station are spelled out. For metering stations covered by standards, specific references to the standards are made. These standards may be government requirements, industry practices, or individual company guidelines; usually they are combinations of these, and detail the kind of meters to be used plus related correcting and readout systems. Details of access by both parties to the equipment and the requirements for frequency of testing and/or reports are spelled out. For large dollar-exchange quantities, there may be an allowance for a check station, with the same provisions listed as above; stating how any discrepancies between the two measurements will be handled (Figure 3-2).

Some provision is made for estimating deliveries during times when the meter is out of service or registering inaccurately, and the procedure for resolving quantities during these periods will be included. Some accuracy limits are set; if these are not met as determined by test or check meters, settlement provisions are implemented. This accuracy limit may typically

Figure 3-2 Typical meter and regulation station components.

be from ±0.5 to ±2.0%, but may be set closer or wider depending on the economics and measurement ability of the meters.

A time period during which a correction can be made is stated if it is not possible to determine the error source and the time of change. This is normally restricted to one half of the time since the last calibration. Requirements for retaining records and reports are spelled out for both parties. This relates to the specified time period allowed for the quantity measurement to be questioned, if major errors are found covering volumes over many months or even years.

Measurements

This specifies, in non-confusing terms, the unit of quantity to be delivered. In a volume measurement, base conditions of temperature and pressure are clearly defined. In a weight (mass) measurement, only the unit of weight need be specified. For most commercial purposes, the terms "weight" and "mass" are used interchangeably without concern about the effects of the attraction of gravity on the weight being measured. Requirements are specified for all related equipment (beyond the basic meter) and how these secondary measurements will be used to correct basic meter readings. These requirements are particularly important, since interested government parties and the parties to the contract in their plant quantity reports may record data on a different base calculation. Major confusion can arise if all of these requirements are not spelled out and clearly understood so that volumes are given on the same basis.

Material Quality

Any natural or manufactured product can have small and varying amounts of foreign material that are not desirable, or at least whose quantities must be limited. The quality section defines the rights of the buyer and the seller if such limits are exceeded. These specifications may also include separate pricing for mixed streams, so quantities must be delineated for proper payment. If there are too many unwanted contaminants, a price reduction may be allowed rather than curtailing a delivery. These details are spelled out in the quality requirement section.

Billing

This section sets a deadline for the computation of quantity, with a provision for correcting errors. It specifies the procedure for billing, the payment period, and penalties for late payment.

Summary of Contract Requirements

A properly written contract, which protects the interests of the buyer and the seller to allow fair and equitable billings to be made for an exchange of the quantity of material, is a basic requirement for establishing custody transfer. The ultimate definition of measurement accuracy is achieved when the seller sends a bill, the buyer pays the bill, and both parties are satisfied with the results. All possible misunderstandings and means of their solution should be defined in the contract in case there is a disagreement.

Other Factors in Custody Transfer

Accuracy

A term used frequently in flow measurement is "accuracy." Accuracy is a term more often abused than correctly used. Unfortunately, it is a sales tool used commercially by both suppliers and users of metering equipment. The supplier with the "best" number wins the bid. Likewise, the user will sometimes require accuracies beyond the capabilities of any meter available. In either case, the accuracy definition serves a purpose for the practitioner or supplier, but has little relevance otherwise. In custody transfer measurement, accuracy is usually defined as the difference between the measured value and the true (reference) value expressed as a percentage. The problem with this definition is that the indicated value is read from the meter, but the method of obtaining the true (reference) value may not be specified; therefore, the true (reference) value is not precisely known. For this reason, it is the subject of many disagreements.

The related term of the "uncertainty" in a specified procedure is a statistical statement with at least a comparative meaning when examining various meter capabilities.

Uncertainty

The performance of the measurement under flowing conditions can be evaluated by making an uncertainty calculation. Many calculation procedures are available in the standards and flow measurement literature. One is "Measurement Uncertainty for Fluid Flow in Closed Conduits" ANSI/ASME MFC-2M. Another is the American Gas Association (AGA) Report No. 3, "Orifice Metering of Natural Gas, Part 1, General Equations and Uncertainty Guidelines." The value of this calculation is not a numerically "absolute" or explicit value (Figure 3-3).

An uncertainty calculation provides an envelope in which the measurement practitioner should expect to find a given percentage of the measured values. In general, that percentage is statistically stated as 2δ, or roughly 95% of the measurements. One of the values of the uncertainty calculation is in examining the significance of each variable that impacts the flow calculation and relating these to the flow measurement application in question. Some of its other values are:

- Engineering departments can use uncertainty calculation information to select the type of equipment to be used in a meter station. Equipment can be selected to meet a system balance expectation or uncertainty.
- It can be used for contract and/or regulatory compliance.
- It can be used by dispatching departments to estimate when the accuracy of a meter station's measurement is changing.
- It can be used to help manage lost and unaccounted for numbers. If equipment that is all one type is installed on the inlet, but is all of another type on the outlet, the metering system may not produce the desired system balance results.
- It can be used by maintenance to understand on which pieces of equipment to concentrate their efforts.

These calculations must consider the particular operating conditions for the specific meter application in order to be most useful in getting the most accurate measurement.

Calculation of the equation's variables is not the whole concern for complete uncertainty determination; allowance must be made for errors in human interpretation, recorders or

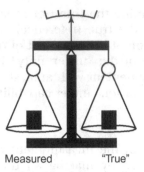

Measured   "True"

Figure 3-3 "Accuracy" in custody transfer means the percentage difference between measured and "true" (reference) flow.

computers, installation, and also for fluid characteristics. However, most of these are assumed to be minimized, provided that industry standard requirements are met and properly trained personnel are responsible for operating and maintaining the station. Since these effects cannot be quantified, they are minimized by recognizing their potential existence and properly controlling meter station design, operation, and maintenance. Without proper attention to the total problem, a simple calculation of the variables in the equation may mislead a practitioner into believing measurement is better or worse than it actually is.

The uncertainty calculation assumes that the meter has been properly installed, operated, and maintained. If maintenance is neglected and the meter has deposits on it that change its flow characteristics, then the calculation is meaningless (Figure 3-4).

Maintenance of Meter Equipment

Both the supplier and the customer must have confidence that a billing meter is reading the proper delivery volumes. Equipment calibration may change over a period of time, so both parties should take an active part in periodically testing the meter system. Without tests to reconfirm original accuracies of the metering system, any statement of accuracy is not complete.

Proving Meters

If there is a desire to reduce measurement tolerances, then an actual throughput test can be run against a "master meter" or a prover system. The master meter should be calibrated and

Figure 3-4 A metering station must be properly installed and maintained in order to provide accurate measurement.

certified to some accuracy limit by a testing facility, a government agency, a private laboratory, a manufacturer, or the practitioner, using agreed-upon flow standards. Periodically, the master meter has to be sent back to the testing facility for recertification. The retesting frequency depends on the fluids being tested and the treatment of the master meter between tests. The best throughput test is one that can be run directly in series with a "prover."

The prover can come in many forms, but essentially it involves a basic volume that has been certified by a government or industry group. Since it is one step closer to a basic calibration, this is the most accurate test of a meter's throughput. Such provers for liquid may be calibrated seraphin cans (for fluids with no vapor pressure at flowing temperature), pressurized volume tanks (for fluids with vapor pressure at the flow temperature), or pipe provers (formerly called mechanical displacement provers as described in the API *Manual of Petroleum Measurement Standards*). Such pipe provers may be permanently installed in large-dollar-volume meter stations, or may be portable units for smaller multiple meter stations (Figures 3-5, 3-6).

Proper maintenance and calibration of a billing meter is essential for accurate custody transfer metering. Testing requires that both the supplier and the customer participate. Diagnostics and evaluation with proper test equipment ensure that recorded volumes are correct. Any proving must be documented and signed by both parties, so that contract provisions can be implemented for any corrections required.

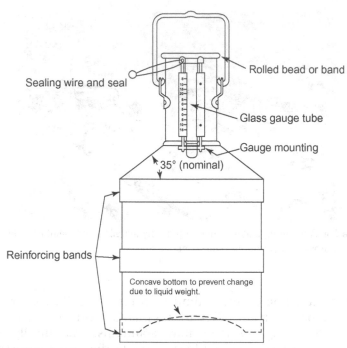

Sealing wire and seal

Rolled bead or band

Glass gauge tube

Gauge mounting

35° (nominal)

Reinforcing bands

Concave bottom to prevent change due to liquid weight.

Figure 3-5 Typical seraphin can prover.

Figure 3-6 Typical pipe prover for liquid meters.

Properly trained personnel, who understand the importance of the equipment they maintain, are the key to accurate measurement. With proper test procedures, accurate test equipment, a good maintenance procedure, and a timely test frequency, any company should have an acceptable "lost or unaccounted for" record.

See Chapter 9 on maintenance for further discussion of this vital phase of instrumentation.

Operation Considerations

Recognition of a meter's operating limits must be considered for a meter to meet its stated accuracy. Most meters operate within stated limits and should not be used in the extremes of ranges for custody transfer metering.

Flow variations should be minimized by better control of the flow rate. If there is no single meter that has the range required to operate in the accurate part of its range, then the use of multiple meters, along with some type of flow switching control to turn meters in and out of service, is required.

In addition to basic meter problems at meter extremes, the secondary equipment that measures pressure, temperature, differential pressure (for head meters), density or specific gravity, and composition of the flow can also have problems. Normal specifications of these devices are stated as percentage of full scale. Selecting an instrument with the wrong range for the parameters to be measured introduces errors (Figure 3-7).

A properly chosen, designed, and installed system may still fail to meet expectations if the meter is not operated in its most accurate range.

Custody Transfer Auditing

When money is exchanged for measured material, the two parties' agreement will include a means of auditing the volumes obtained. Sufficient operation and maintenance records made available to both parties will ensure that the calculated volumes can be arrived at independently. At least a check of the values used by the other party should be made to ensure that agreement is reached over the volume.

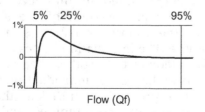

Figure 3-7 For best accuracy, operate this meter above about 25% and below about 95% of full scale.

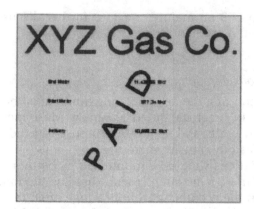
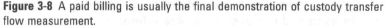

Figure 3-8 A paid billing is usually the final demonstration of custody transfer flow measurement.

This procedure is an important aspect of custody transfer metering, and is usually completed within 30 to 60 days after a bill is submitted. It will keep both parties involved in the measurement, and will prevent disagreements about procedures and volumes at some later date. With the data still current, a disagreement can be settled while knowledge of the measurement is fresh in both parties' minds. A complete file of any disagreements must be kept, including resolutions. Records can be reviewed to see whether a particular station or particular errors are involved in recurring problems that need to be addressed by an equipment or maintenance upgrade (Figure 3-8).

Summary of Custody Transfer

Custody transfer measurement begins with a contract between two parties that specifies the data needed to choose a metering system. To get the most accurate (minimum uncertainty) measurement required to minimize settlement problems, maintenance and operation of the system must be controlled so that the accuracy capabilities of the meter may be realized in service. The information in this chapter should be supplemented by reference to the other chapters of the book for a complete understanding of an individual meter's advantages and limitations. If all precautions are taken, then proof of the station will be when bills are submitted and paid, and the custody has been successfully transferred.

Non-Custody Transfer Measurement

Set Point Control Signal

Of prime importance in any process is the ability to measure flows so that the process can be controlled. The absolute accuracy of such a signal is not as important as the ability to repeat the measurement under the same flow conditions. It is confusing to speak of not needing accuracy for flow signals as much as needing repeatability, but seldom does a process operate at its original design parameters. The process must be tuned before it comes into balance. Then the function of the set point control signal is to hold the balance and make the changes required when the process is varied.

An unpredictable meter output can cause control problems, so a control signal must come from a repeatable measurement at given rates across the range of flow. Because of the sensitivity of some processes to changes, the response time of a flow control signal is much more critical than for a custody transfer. In set point control measurement, the rate signal represents the process variable of interest, whereas in custody transfer, the total flow is required. However, in contrast to custody transfer measurement where the ability to integrate volume over time is a significant part of the system, the set point control signal is seldom integrated to totalize the volume.

Other Uses

Flow may be measured periodically to check an operation with the assumption that it will then run properly until results indicate otherwise. A good example is a heating and cooling distribution system using ducts. Once set, the system is changed only if there is an indication that the distribution has changed as heating or cooling gets out of control. Similar tests are done for pollution studies. In these "other uses," accuracy may be no better than ±5 to ±10%.

Flow may also be used as an alarm signal to indicate that the design limits of the flow volumes (either high or low) have been exceeded, and action should be taken.

Summary of Flow Measurement

Many different capabilities are required to measure flow. Each application should be defined so that expectations of accuracy can be balanced against cost to derive the most cost-effective installation that will do what is required.

References

American Petroleum Institute. Manual of Petroleum Measurement Standards, Washington, DC.

ANSI/ASME. MFC-2M: Measurement Uncertainty for Fluid Flow in Closed Conduits. Society of Mechanical Engineers, New York.

BASIC REFERENCE STANDARDS

Flow has been measured since the earliest of times. For example, flow measurement was used to control the taxation of landowners in the Nile delta in ancient times. Deposits of rich soil occurred each year during flooding. The amount of land flooded was related to the height of the water level on a calibrated stick whose calibration was under the control of the king. At the height of the flood, the stick would be read and taxes for the coming year set on the basis of how much land would be enriched by the flood. When budgets got a little tight, the king simply recalibrated the stick!

The standards for measurement over the years represent the continuous upgrading of knowledge on older meters and establishment of new standards for newly developed meters. Some of the organizations that have been involved in writing these standards are: the American Petroleum Institute (API), the American Gas Association (AGA), the Society of Petroleum Engineers (SPE), the American Society of Mechanical Engineers (ASME), the Gas Processors Association (GPA), the Instrument Society of America (ISA), the American National Standards Institute (ANSI), the American Society for Testing and Materials (ASTM), the Institute of Petroleum (IP), the British Standards Institution (BSI), the International Organization for Standardization (ISO), the International Organization of Legal Metrology (OIML), and the European Community (EC).

In addition to these standards organizations, much published and user data are available from various universities, measurement schools, manufacturers, industry organizations, governmental agencies, books, and individuals. Individuals represent a particularly useful source of information about the metering of specialty fluids that may or may not be included in a standard. For new meters, information from the manufacturer must be used initially because new meters have only limited use; detailed industry information is scarce (Figure 4-1).

Fluid Flow Measurement. ISBN: 978-0-12-409524-3

Figure 4-1 A typical collection of standards.

Once a meter is widely used and develops a track record, standards are written. Specific examples of standards are listed below.

Most organizations involved in flow measurement, such as ANSI, API etc., require that standards are reviewed once every five years and either reconfirmed or updated based on the responsible committee's action. Therefore, it is important to check with the organization to ensure that the latest or applicable version is obtained and used.

The website address listed after each organization can be used to check on the latest standard or reports available.

American Gas Association (AGA)

1515 Wilson Boulevard
Arlington, VA 22209
Phone (toll free): (866) 816–9444
Fax: (201) 986–7886
Order: ILI Infodisk, Inc.
610 Winters Avenue
Paramus, NJ 07862
Web site: www.aga.org

Engineering and Operations: Measurement

AGA Report No. 3, "Orifice Metering of Natural Gas"
Note: AGA-3 is listed under the API Publications as Chapter 14, "Natural Gas Fluids Measurement." Both

organizations publish identical reports, any of which may be ordered from either with a price advantage to members.

AGA Report No. 3, Part 1, "General Equations & Uncertainty Guidelines"

Part 1 provides the basic equations and uncertainty statements for computing the flow through orifice meters.

AGA Report No. 3, Part 2, "Specification and Installation Requirements"

Part 2 provides the specifications for construction and installation of orifice plates, meter tubes, and associated fittings.

AGA Report No. 3, Part 3, "Natural Gas Applications"

Part 3 provides practical guidelines for the measurement of natural gas. Mass flow rate and volumetric rate methods are presented in conformance with North American industry practices.

AGA Report No. 3, Part 4, "Background, Development Implementation Procedure"

Part 4 provides instruction on implementation, including subroutine documentation. It also explains the historical development of the revised standard.

AGA Report No. 4A, "Natural Gas Contract, Measurement and Quality Clauses"

For use by gas industry measurement engineers and legal personnel, this report provides guidelines for custody transfer contracts and discusses the three gas measurement concepts: accounting for gas in units of volume, energy or weight.

AGA Report No. 5, "Fuel Gas Energy Metering"

A supplement to published measurement procedures, this publication provides for the conversion of units of gas volume or mass-to-energy equivalents through the use of data associated with volume-metering practices.

AGA Report No. 7, "Measurement of Gas by Turbine Meters"

Provides information on the theory of operation, performance characteristics, and installation and maintenance of turbine meters. Also includes techniques for flow computation, calibration, and field checks.

AGA Report No. 8, "Compressibility and Supercompressibility for Natural Gas and Other Hydrocarbon Gases"

Presents information needed (including efficient FORTRAN 77 computer program listings) to compute gas phase densities and compressibility and supercompressibility factors for natural gas and other related hydrocarbon gases.

AGA Report No. 9, "Measurement of Natural Gas by Multipath Ultrasonic Meters"

This report is for multipath ultrasonic transit-time flow meters, typically 6 inches and larger in diameter, used for the measurement of natural gas. It is written in the form of a performance-based specification.

ANSI, B109.1–B109.4 for Diaphragm-Type Gas Displacement and Rotary Meters, and Gas Measurement Manual Set: Parts 1 to 15

A set of 15 indispensable references on gas measurement, including practices, calculations, theory, and history. Also sold separately.

"Gas Orifice Flow Program" based on AGA Reports No. 3 and No. 8

There are nine separate programs of various combinations of computers and formats. Contact the AGA for a full listing (Figure 4-2).

Figure 4-2 Some of the many volumes in the API *Manual of Petroleum Measurement Standards*.

American Petroleum Institute (API)

Ordering address:
American Petroleum Institute
Publications and Distribution
1220 L St, NW
Washington, DC 20005
Phone: (202) 682–8375
website: www.api.org

The API maintains a comprehensive "Manual of Petroleum Measurement Standards." This manual is an ongoing project, as new chapters and sections of old chapters are released periodically. Listed here are the chapters pertinent to flow measurement, specifically as it relates to petroleum products. (Note: These were the latest editions at the time of the publication of this book [2014].)

Chapter 1, "Vocabulary." This chapter defines and describes the words and terms used throughout the manual.

Chapter 4, "Proving Systems." This chapter serves as a guide for the design, installation, calibration, and operation of meter proving systems.

Chapter 4.1, "Introduction." This chapter is a general introduction to the subject of proving, the procedure used to determine a meter factor.

Chapter 4.2, "Conventional Pipe Provers." This chapter outlines the essential elements of unidirectional and bidirectional conventional pipe provers and provides design, installation, and calibration details for the types of pipe provers currently in use.

Chapter 4.3, "Small Volume Provers." This chapter outlines the essential elements of a small volume prover and provides descriptions of, and operating details for, the various types of small volume provers that meet acceptable standards of repeatability and accuracy.

Chapter 4.4, "Tank Provers." This chapter specifies characteristics of tank provers that are in general use and the procedures for their calibration. This publication does not apply to weir-type, vapor-condensing dual-tank water-displacement, or gas-displacement provers.

Chapter 4.5, "Master-Meter Provers." This chapter covers the use of both displacement and turbine meters as master meters.

Chapter 4.6, "Pulse Interpolation." This chapter describes how the double-chronometry method of pulse interpolation, including system operating requirements and equipment testing, is applied to meter proving.

Chapter 4.7, "Field-Standard Test Measures." This chapter outlines the essential elements of field-standard test measures and provides descriptions and operating details. The volume range of measures in this chapter is 1 to 1,500 gallons.

Chapter 5, "Metering." This chapter covers the dynamic measurement of liquid hydrocarbons, or metering. It is divided into subchapters as follows:

Chapter 5.1, "General Considerations for Measurement by Meters." This chapter is an overall introduction to Chapter 5, "Metering."

Chapter 5.2, "Measurement of Liquid Hydrocarbons by Displacement Meters." This chapter describes methods of obtaining accurate measurements and maximum service life when displacement meters are used to measure liquid hydrocarbons.

Chapter 5.3, "Measurement of Liquid Hydrocarbons by Turbine Meters." This chapter defines the application criteria for turbine meters and discusses appropriate considerations regarding the liquids to be measured, the installation of turbine metering systems, and the performance, operation, and maintenance of turbine meters in liquid hydrocarbon service.

Chapter 5.4, "Accessory Equipment for Liquid Meters." This chapter describes characteristics of accessory equipment that is generally used with displacement and turbine meters in liquid hydrocarbon service.

Chapter 5.5, "Fidelity and Security of Flow Measurement Pulsed-Data Transmission Systems." This chapter provides a guide to the selection, operation, and maintenance of pulse-data, cabled transmission systems for fluid metering systems to provide the desired level of fidelity and security of transmitted data.

Chapter 6, "Metering Assemblies." This chapter discusses the design, installation, and operation of metering systems for coping with special situations in hydrocarbon measurement.

Chapter 6.1, "LACT Systems." This chapter serves as a guide for the design, installation, calibration, and operation of lease automatic custody transfer systems.

Chapter 6.2, "Loading Rack and Tank Truck Metering Systems for Non-LPG Products." This chapter guides the selection and installation of loading rack and tank truck metering systems for most gasoline and oil products other than liquefied petroleum gas.

Chapter 6.3, "Service Station Dispensing Metering Systems." This chapter covers service station metering systems used for dispensing motor fuel (except liquefied petroleum gas fuels) to road vehicles at relatively low flow and pressure.

Chapter 6.4, "Metering Systems for Aviation Fueling Facilities." This chapter is a guide to the selection, installation, performance, and maintenance of metering systems for aviation dispensing systems.

Chapter 6.5, "Metering Systems for Loading and Unloading Marine Carriers." This chapter deals with the operation and special arrangement of meters, provers, manifolding, instrumentation, and accessory equipment used for measurement in loading and unloading marine bulk carriers.

Chapter 6.6, "Pipeline Metering Systems." This chapter provides guidelines for selection of type and size of measurement pipeline oil movements, as well as the relative advantages and disadvantages of three meter proving methods.

Chapter 6.7, "Metering Viscous Hydrocarbons." This publication serves as a guide for the design, installation, operation, and proving of meters and their auxiliary equipment used to meter viscous hydrocarbons.

Chapter 7, "Temperature Determination." This chapter covers the sampling, reading, averaging, and rounding of the temperature of liquid hydrocarbons in both the static and dynamic modes of measurement for volumetric purposes. The following chapters and standards now cover the subject of temperature determination.

Chapter 7.2, "Dynamic Temperature Determination." This section describes the methods and practices used to obtain flowing temperature using portable electronic thermometers in custody transfer of liquid hydrocarbons.

Chapter 7.3, "Static Temperature Determination Using Portable Electronic Thermometers." This section provides a guide to the use of portable electronic thermometers to determine temperatures for custody transfer of liquid hydrocarbons under static conditions.

Chapter 8, "Sampling." This chapter covers standardized procedures for sampling crude oil or its products. It is divided into subchapters as follows:

Chapter 8.1, "Manual Sampling of Petroleum and Petroleum Products." This chapter covers the procedures for obtaining representative samples of shipments of uniform petroleum products, except electrical insulating oils and fluid power hydraulic fluids. It also covers sampling of crude petroleum and non-uniform petroleum products and shipments. It does not cover butane, propane, and gas liquids with a Reid Vapor Pressure (RVP) above 26.

Chapter 8.2, "Automatic Sampling of Petroleum and Petroleum Products." This chapter covers automatic procedures for obtaining representative samples of petroleum and non-uniform stocks for shipments, except electrical insulating oil.

Chapter 9, "Density Determination." This chapter, which describes the standard methods and apparatus used to determine specific gravity of crude petroleum products normally handled as liquids, is divided into subchapters as follows:

Chapter 9.1, "Hydrometer Test Method for Density, Relative Density (Specific Gravity), or API Gravity of Crude Petroleum and Liquid Petroleum Products." This chapter describes the methods and practices relating to the determination of the density, relative density, or API gravity of crude petroleum and liquid petroleum products using the hydrometer method (laboratory determination).

Chapter 9.2, "Pressure Hydrometer Test Method for Density, Relative Density." This chapter provides a guide for determining the density or relative density (specific gravity) or API gravity of light hydrocarbons, including liquefied petroleum gases, using the pressure hydrometer.

Determination of Water and Sediment

Chapter 10, "Sediment and Water." This chapter describes methods for determining the amount of sediment and water, either together or separately. Laboratory and field methods are covered as follows:

Chapter 10.1, "Determination of Sediment in Crude Oils and Fuel Oils Extraction Method." This publication specifies a method for the determination of sediment in crude petroleum by extraction with toluene.

Chapter 10.2, "Determination of Water in Crude Oil by Distillation." This publication specifies a method for the determination of sediment in crude petroleum by distillation.

Chapter 10.3, "Determination of Water and Sediment in Crude Oil Centrifuge Method (Laboratory Procedure)." This publication describes the method of laboratory determination of water sediment in crude oil by means of the centrifuge procedure.

Chapter 10.4, "Determination of Sediment and Water in Crude Oil Centrifuge Method (Field Procedure)." This chapter describes procedures for the determination of water and sediment in crude oils using the field centrifuge procedure.

Chapter 10.5, "Determination of Water in Petroleum Products and Bituminous Materials by Distillation." This publication describes the test method for the determination of water levels in petroleum products and bituminous materials by distillation.

Chapter 10.6, "Determination of Water and Sediment in Fuel Oils by the Centrifuge Method (Laboratory Procedure)." This publication describes the test method for laboratory determination of water and sediment in fuel oils by centrifuge.

Chapter 10.7, "Standard Test Method for Water in Crude Oil by Karl Fischer Titration (Potentiometric)." This covers the determination of water in the range from 0.02 to 2 mass percent in crude oil.

Chapter 10.8, "Standard Test Method for Sediment in Crude Oil by Membrane Filtration." This method has been validated for crude oils with sediment content of approximately 0.15 mass percent.

Chapter 10.9, "Standard Test Method for Water in Crude Oil by Coulometric Karl Fischer Titration."

Basic Calculation Data

Chapter 11, "Physical Properties Data." Because of the nature of this material, it is not included in the complete set of measurement standards. Each element of Chapter 11 must be ordered separately. Chapter 11 contains the physical data that have direct application to volumetric measurement of liquid hydrocarbons. These are presented in tabular form, in equations relating volume to temperature and pressure, computer subroutines, and magnetic tape.

Chapter 11.1, Vol. 1:
> Table 5A, "Generalized Oils and JP-4, Correction of Observed API Gravity to API Gravity at 60°F."

> Table 5B, "Generalized Products, Correction of Observed API Gravity to API Gravity at 60°F."

> Table 6A, "Generalized Crude Oils and JP-4, Correction of Volume to 60°F Against API Gravity at 60°F."

Chapter 11.1, Vol. 2:
> Table 6B, "Generalized Products, Correction of Volume to 60°F Against API Gravity at 60°F."

Chapter 11.1, Vol. 3:
> Table 6C, "Volume Correction Factors for Individual, and Special Applications, Correction to 60°F Against Thermal Expansion Coefficients at 60° F."

Chapter 11.1, Vol. 4:
> Table 23A, "Generalized Crude Oils, Correction of Observed Relative Density to Relative Density at 60/60°F.

> Table 24A, "Generalized Crude Oil, Correction of Volume to 60°F Against Relative Density 60/60°F."

Chapter 11.1, Vol. 5:
> Table 23B, "Generalized Products, Correction of Observed Relative Density to Relative Density at 60/60°F."

Chapter 11.1, Vol. 6:
> Table 24C, "Volume Correction Factors for Individual and Special Applications, Volume
> Correction to 60°F Against Thermal Expansion Coefficients at 60°F."

Chapter 11.1, Vol. 7 :
> Table 53A, "Generalized Crude Oils, Correction of Observed Density to Density at 15°C."

Table 54A "Generalized Crude Oils, Correction of Volume to 15°C Against Density at 15°C."

Chapter 11.1, Vol. 8:

Table 53B, "Generalized Products, Correction of Observed Density to Density at 15°C."

Table 54B, "Generalized Products, Correction of Volume to 15°C Against Density at 15°C."

Chapter 11.1, Vol. 9 (reaffirmed March 1997):

Table 54C, "Volume Correction Factors for Individual and Special Applications, Volume Correction to 15°C Against Thermal Expansion Coefficients at 15°C."

Chapter 11.1, Vol. 10: Background, development, and computer documentation, including computer subroutines in Fortran IV for all volumes of Chapter 11.1 except Volumes 11, 12, 13, and 14. Implementation procedures, including rounding and truncating procedures, are also included. These subroutines are not available through API in magnetic or electronic form.

Chapter 11.2.1, "Compressibility Factors for Hydrocarbons: 0–90° API Gravity Range." This chapter provides tables to correct hydrocarbon volumes metered under pressure to corresponding volumes at the equilibrium pressure for the metered temperature. It contains compressibility factors related to meter temperature and API gravity (60°F) of metered material.

Chapter 11.2.2, "Compressibility Factors for Hydrocarbons: 0.350–0.637 Relative Density (60°F/60°F) and 50°F to 140°F Metering Temperature." This publication provides tables to correct hydrocarbon volumes metered under pressure to corresponding volumes at equilibrium pressure for the metered temperature. The standard contains compressibility factors related to the meter temperature and relative density (60°F/60°F) of the metered material.

Chapter 11.2.3, "Water Calibration of Volumetric Provers." This chapter contains volume correction factors in standard units related to prover temperature, and the difference in temperature between the prover and a certified test measure.

Chapter 11.3.2.1, "Ethylene Density." This chapter is a computer tape that will produce either a density (pounds/ft^3) or a compressibility factor for vapor phase ethylene over the temperature range from 65° to 167°F and the pressure range from 200 to 2,100 psia. The tape is 9-track, 1,600 bpi, unlabeled, and is available in either ASCII or EBCDIC. The desired format must be specified when ordering.

Chapter 11.3.3.2, "Propylene Compressibility." This chapter is a computer tape that will produce a table of values applicable to liquid propylene in the following ranges: temperatures 30° to 165°F and saturation pressure to 1,600 psia. The tape computes the following two values: density (pounds/ft^3) at flowing temperature and temperature, and ratio of density at flowing conditions to density at 60°F and saturation pressure. The tape is 9-track, 1,600 bpi, unlabeled, and is available in either ASCII or EBCDIC format. The desired format must be specified when ordering.

Flow Calculation Procedures

Chapter 12, "Calculation of Petroleum Quantities." This chapter describes standard procedures for calculating net standard volumes, including the application of correction factors and the importance of significant figures. The purpose of standardizing the calculation procedure is to achieve the same result regardless of what person or computer does the calculating.

Chapter 12.2, "Calculation of Liquid Petroleum Quantities Measured by Turbine or Displacement Meters." This publication defines the terms used in the calculation of metered petroleum properties, and specifies the equations that allow values of correction factors to be computed. The rules for sequence, rounding, and significant figures are provided, along with tables for computer calculations.

Chapter 12.3, "Calibration of Volumetric Shrinkage from Blending Light Hydrocarbons with Crude Oils." This publication presents data on the subject of volumetric changes resulting from blending volatile hydrocarbons (propane, butane, produced distillates, and natural gasolines) with crude oils. This publication is not included in the current manual.

Chapter 13, "Statistical Aspects of Measuring and Sampling." The more accurate petroleum measurement becomes, the more its practitioners stand in need of statistical methods to express residual uncertainties. This chapter covers the application of statistical methods to petroleum measurement and sampling. Chapter 13 is in preparation. The following portion now covers statistical aspects of measuring and sampling and is included in the manual.

Chapter 13.1, "Statistical Concepts and Procedures in Measurement." This chapter is designed to help those who make measurement of bulk oil quantities improve the value of their result statement by making proper estimates of the uncertainty or probable error involved in measurements.

Chapter 14, "Natural Gas Fluids Measurement." This chapter standardizes practices for measuring, sampling, and testing natural gas fluids. Chapter 14 is in preparation. Sections 3, 5, 6, and 8 have been completed and are included in the manual.

Chapter 14.1, "Collecting and Handling of Natural Gas Samples for Custody Transfer."

Chapter 14.3, Part 1, "General Equations and Uncertainty Guidelines." Part 1 provides basic equations and uncertainty statements for computing flow through orifice meters.

Chapter 14.3, Part 2, "Specifications and Installation Requirements."

Chapter 14.3, Part 3, "Natural Gas Applications."

Chapter 14.3, Part 4, "Background, Development, Implementation Procedures, and Subroutine Documentation."

Chapter 14.4, "Converting Mass of Natural Gas Liquids and Vapors to Equivalent Liquid Problems."

Chapter 14.5, "Calculation of Gross Heating Value, Specific Gravity, and Compressibility of Natural Gas Mixtures from Compositional Analysis." Outlines procedures to calculate, from compositional analysis, the following properties of natural gas mixtures: heating value, specific gravity, and compressibility factor.

Chapter 14.6, "Continuous Density Measurement." Formerly titled "Installing and Proving Density Meters," this chapter provides criteria and procedures for designing, operating, and calibrating continuous density measurement systems for Newtonian fluids in the petroleum, chemical, and natural gas industries. The application of this standard is limited to clean, homogeneous, single phase fluids or supercritical fluids whose flowing density is greater than 0.3 grams per cubic centimeter at operating conditions of 60°F (15.6°C) and saturation pressure.

Chapter 14.8, "Liquefied Petroleum Gas Measurement." This chapter describes dynamic and static metering systems used to measure liquefied petroleum gas in the density range of 0.30 to 0.70 grams per cubic centimeter.

Chapter 15, "Guidelines for Use of the International System of Units (SI) in the Petroleum and Allied Industries." This publication specifies the API-preferred units for quantities involved in petroleum industry measurements and indicates factors for conversion of quantities expressed in customary units to the API-preferred metric units. The quantities that comprise the tables are grouped into convenient categories related to their use. They were chosen to meet the needs of the many and varied aspects of the petroleum industry but also should be useful in similar process industries.

Chapter 16, "Measurement of Hydrocarbon Fluids by Weight or Mass." This chapter covers the static and dynamic measurement of hydrocarbon fluids by weight or mass.

Chapter 17, "Marine Measurement." This chapter provides guidelines for the measurement and reporting of crude oil or petroleum product transfers by share terminal operators, vessel personnel, other parties involved in terminal marine cargo transfer measurement, and accounting operations.

Chapter 18, "Custody Transfer." This chapter covers application of other measurement standards to unique custody transfer situations (production measurement).

Chapter 19, "Evaporation Loss Measurement." Marine vessels, low pressure tanks, pressure vacuum vent valves and other vessels.

Chapter 20, "Allocation of Measurement of Oil and Natural Gas." Production measurement design and operating guidelines for liquid and gas allocation measurement systems.

Recommendations for metering, static measurement, sampling, proving, calibrating, and calculation procedures.

Chapter 21, "Flow Measurement Using Electronic Metering Systems." Defines standard practices and minimum specifications for electronic measurement systems used in the measurement and recording flow parameters. This chapter covers natural gas fluid and petroleum, and petroleum product custody transfer applications using industry-recognized primary measurement devices (Figures 4-3 and 4-4).

Figure 4-3 For many applications, API and AGA standards are the same.

Figure 4-4 Other bodies whose standards impact the flow measurement industry include ASME, ISA, ISO, and others.

American Society of Mechanical Engineers (ASME)

United Engineering Center
345 East 47th St.
New York, NY 10017
Attn: Publications Department
Phone: (202) 822–1167
website: www.asme.org

MFC-IM "Glossary of Terms Used in the Measurement of Fluid Flow in Pipes"

MFC-2M "Measurement Uncertainty for Fluid Flow in Closed Conduits"

MFC-3M "Measurement of Fluid Flow in Pipes Using Orifice, Nozzle, and Venturi" (not an American National Standard)

MFC-4M "Measurement of Gas Flow by Turbine Meters"

MFC-5M (1985, R1994) "Measurement of Liquid Flow in Closed Conduits Using Transit-Time Ultrasonic Flowmeters"

MFC-6M "Measurement of Fluid Flow in Pipes Using Vortex Flow Meters"

MFC-7M "Measurement of Gas Flow by Means of Critical Flow Venturi Nozzles"

MFC-8M "Fluid Flow in Closed Conduits – Connections for Pressure Signal Transmissions between Primary and Secondary Devices"

MFC-9M "Measurement of Liquid Flow in Closed Conduits by Weighing Method"

MFC-10M "Method for Establishing Installation Effects on Flowmeters"

MFC-11M "Measurement of Fluid Flow by Means of Coriolis Mass Flowmeters"

MFC-14M "Measurement of Fluid Flow Using Small Bore Precision Orifice Meters"

MFC-16M "Measurement of Fluid Flow in Closed Conduit by Means of Electromagnetic Flowmeters"

American Society for Testing and Materials (ASTM)

100 Bar Harbor Drive
West Conshohocen, PA 19428-2953

D1070-85 "Standard Test Methods for Relative Density of Gaseous Fuels"

D1072-90 e 1 "Standard Test Method for Total Sulfur in Fuel Gases"

D1142 "Standard Test Method for Water Vapor Content of Gaseous Fuels by Measurement of Dew-Point Temperature"

D1826 "Standard Test Method for Calorific (Heating) Value of Gases in Natural Gas by Continuous Recording Calorimeter"

D1945 "Standard Test Method for Analysis of Natural Gas by Gas Chromatography"

D1988 e 1 "Standard Test Method for Mercaptans in Natural Gas Using Length-of-Stain Detector Tubes"

D3588 "Standard Practice for Calculating Heat Value, Compressibility Factor, and Relative Density of Gaseous Fuels"

D4810 "Standard Test Method for Hydrogen Sulfide in Natural Gas Using Length-of-Stain Detector Tubes"

D4888 "Standard Test Method for Water Vapor in Natural Gas Using Length-of-Stain Detector Tubes"

D4984 "Standard Test Method for Carbon Dioxide in Natural Gas Using Length-of-Stain Detector Tubes"

D5287 "Standard Practice for Automatic Sampling of Gaseous Fuels"

D5454 "Standard Test Method for Water Vapor Content of Gaseous Fuels Using Electronic Moisture Analyzers"

D5503 "Standard Practice for Natural Gas Sample Handling and Conditioning Systems for Pipeline Instrumentation"

Gas Processors Association (GPA)

Patricia Preast
6526 E. 60th Street
Tulsa, OK 74145
Phone: (918) 493–3872
Fax: (918) 493–3875
E-mail: ppreast@gasprocessors.com

CD-ROM: Most GPA documents listed are now available on CD-ROM through:

Information Handling Services (HIS)
15 Inverness Way East
PO Box 1154
Englewood, CO 80150-1154
Phone: (800) 241–7824 or fax: (303) 397–2599

GPA Standard 2165, "Standard for Analysis of Natural Gas Liquid Mixtures by Gas Chromatography"

GPA Standard 2177, "Analysis of Demethanized Hydrocarbon Liquid Mixtures Containing Nitrogen and Carbon Dioxide by Gas Chromatography"

GPA Standard 2186 "Tentative Method for the Extended Analysis of Hydrocarbon Liquid Mixtures Containing Nitrogen and Carbon Dioxide by Temperature Programmed Gas Chromatography"

GPA Standard 2261 "Analysis for Natural Gas and Similar Gaseous Mixtures by Gas Chromatography"

GPA Standard 2265 "GPA Standard for Determination of Hydrogen Sulfide and Mercaptan Sulfur in Natural Gas (Cadmium Sulfate-Iodometric Titration Method)"

GPA Standard 2286 "Tentative Method of Extended Analysis for Natural Gas and Similar Gaseous Mixtures by Temperature Programmed Gas Chromatography"

GPA Standard 2377 "Test for Hydrogen Sulfide and Carbon Dioxide in Natural Gas Using Length of Stain Tubes"

Measurement Standards

GPA Standard 2145, Rev. 2 "Physical Constants for Paraffin Hydrocarbons and Other Components of Natural Gas. Data are given in both English and SI Units"

GPA Standard 2172 "Calculation of Gross Heating Value, Relative Density, and Compressibility of Natural Gas Mixtures from Compositional Analysis"

GPA Standard 8173 "Method for Converting Mass Natural Gas Liquids and Vapors to Equivalent Liquid Volumes" Data are given in both English and SI Units.

GPA Standard 8182 "Standard for the Mass Measurement of Natural Gas Liquids"

GPA Standard 8195 "Tentative Standard for Converting Net Vapor Space Volumes to Equivalent Liquid Volumes"

GPA Reference Bulletin 181 "Heating Value as a Basis for Custody Transfer of Natural Gas" A reference to provide authoritative interpretation of accepted procedures for determining heating values.

GPA Reference Bulletin 194 "Tentative NGL Loading Practices"

Sampling Methods

GPA Standard 2166 "Obtaining Natural Gas Samples for Analysis by Gas Chromatography"

GPA Standard 2174 "Obtaining Liquid Hydrocarbon Samples for Analysis by Gas Chromatography"

Miscellaneous Standards

GPA Publication 1167 "GPA Glossary – Definition of Words and Terms Used in the Gas Processing Industry"

Research Reports

Results of most of the GPA sponsored research projects since 1971 have been published as numbered research reports. All research reports are available upon request.

Instrument Society of America (ISA)

PO Box 12277
Research Triangle Park, NC 27709
Phone (800) 334–6391
website: www.iso.org
17.120 "Measurement of Fluid Flow"

17.120.01 "Measurement of Fluid Flow General"

17.120.10 "Flow in a Closed Conduit"

The following books are available from ISA:

DeCarlo, J.P. 1984. *Fundamentals of Flow Measurement.* Provides a basic working knowledge of the methods of flow measurements.

Spitzer, D.W. 1990. *Industrial Flow Measurement.* 2nd ed. Effective flow meter selection requires a thorough understanding of flow meter technology plus a practical knowledge of the fluid being measured. This resource reviews important flow measurement concepts to help practicing engineers avoid the costs of misapplication. The text provides explanations, practical considerations, illustrations, and examples of existing flow meter methodology. A rational procedure for flow meter selection is presented to help decision makers evaluate appropriate criteria.

The ISA *Recommended Practice Guides* cover flow measurement and related instrumentation and are particularly directed to plant operations (Figure 4-5). A listing of these is available from ISA Publications. In addition to the documents above, ISA offers training courses for self study and also at its facility in North Carolina. Information on these and other services is available in the yearly *Catalogue of Publications and Training Products.*

Spitzer, D.W., ed. 1991. *Flow Measurement.* This is part of the "Practical Guide Series for Measurement and Control."

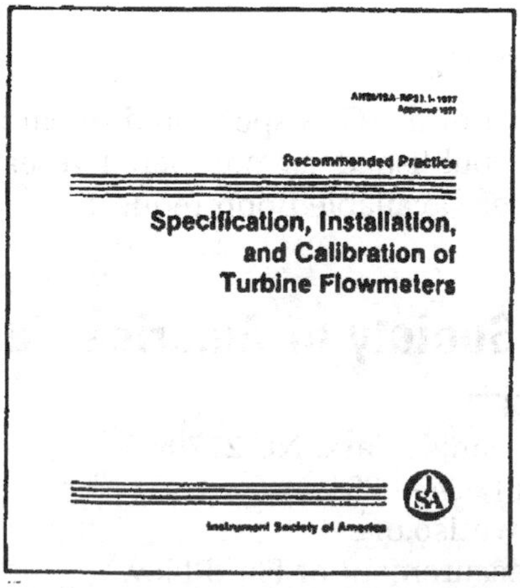

Figure 4-5 Typical Recommended Practice available from ISA.

Addresses for other Sources of Measurement Standards

British Standards Institution
389 Chiswick High Road, LondonW4 4AL
England
Phone: +44 20 8996 9000
website: www.bsigroup.co.uk

International Organization for Standardization (ISO)
1, ch. de la Voie-Creuse
CH-1211 Geneva 20, Switzerland
website: www.iso.org

International Organization of Legal Metrology
Bureau International de Metrologic Legale
11 rue Turgot
75009, Paris, France
Phone +33 1 48 78 12 82

Key Industry Web Links

The listings below are web sites that have related information on the subject of fluid flow measurement. The e-mail addresses can be used to access these sources of information to review the data available.

Fluid Flow Measurement Standards Writing Groups

American Gas Association: www.aga.org
American National Standards Institute: www.ansi.org
American Petroleum Institute: www.api.org
American Society of Mechanical Engineers International: www.asme.org
Gas Industry Standards Board: www.neosoft.com/~gisb/gisb.htm
Gas Processors Association: www.gasprocessors.org
International Organization of Legal Metrology: www.oiml.org
International Organization for Standardization; www.iso.org
National Institute of Standards and Technology: www.nist.gov

Other Standards Writing Groups

Electronic Industries Association: www.eia.org
European Committee for Electrotechnical Standardization: www.cenelec.be

European Committee for Standardization: www.cenorm.be
European Gas Research Group: www.icgti.org/open/organ/gerg
Institute of Electrical and Electronic Engineers, Inc.: www.ieee.org
International Electrotechnical Commission: www.iec.ch/home-e.htm
Underwriters Laboratories, Inc.: www.ul.com

International Energy Organizations

International Energy Agency: www.iea.org
International Center for Gas Technology Information: www.icgti.org
International Gas Union: www.igu.org
PRC International: www.prci.com
Society of Petroleum Engineers: www.spe.org
World Energy Council: www.worldenergy.org

United States Government

Bureau of Land Management: www.blm.gov
Department of Energy: www.doe.gov
Department of Transportation: www.dot.gov
Energy Information Administration: www.eia.doe.gov
Environmental Protection Agency: www.epa.gov
Federal Energy Regulatory Commission: www.ferc.fed.us
Federal Energy Technology Center: www.fetc.doe.gov
Minerals Management Service: www.mms.gov
National Petroleum Technology Office: www.npto.doe.gov
Occupational Safety & Health Administration: www.osha.gov
Petroleum Technology Transfer Council: www.pttc.org

National Fossil Fuel Industry Organizations

American Institute of Chemical Engineers: www.aiche.org
American Public Gas Association: www.apga.org
Gas Machinery Council: www.gmrc.org
Gas Research Institute: www.gri.org
Independent Petroleum Association of America: www.ipaa.org
National Association of Regulatory Utility Commissioners: www.naruc.org
Natural Gas Information and Educational Resource: www.naturalgas.org

Regional Fossil Fuel Industry Organizations

Independent 4 and Natural Gas Association:
www.wvonga.com

Industry Trade Publications

Gas Utility and Pipeline Industries: www.gasindustries.com
American Gas: www.aga.com/magazine/amgas.html
COMPRESSORtech^2: www.dieselpub.com
Flow Control Magazine: www.flowcontrolnetwork.com
Gas Daily's NG: www.ftenergyusa.com/gasdaily
Pipeline & Gas Industry: www.pipe-line.com
Pipeline & Gas Journal: www.undergroundinfo.com
Oil & Gas Journal: www.ogjonline.com
World Oil: www.worldoil.com
Hart Publications: www.hartpub.com

Other Energy Industry Links

EnergyOnline: www.energyonline.com
EnergySource.com: www.energysource.com
Energy Research Clearing House: www.main.com/~ERCH/
Gas.com: www.gas.com
Institute of Energy: www.entech.co.uk/ioe/ioepro.htm
Naturalgas.com: www.naturalgas.com
Utility Connection: www.magicnet.net/~metzler/

During the reading of this chapter, the reader may have noticed that many of the reference standards from different organizations address the same subject matter. The reader is cautioned that even though this is the case, the documents from different organizations do not necessarily provide the same recommendation and calculation results. For example, for orifice meter measurement one will find three different calculation results and two different installation requirements. One might ask why. The answer is that the documents were generated by different groups (API ASME, ISO, and OIML) and may represent different revisions or were generated with:

- A lack of common scope.
- A lack of consensus on research criteria.
- A lack of definition of meaningful difference.

Numerous attempts to reconcile the differences have occurred with very little success due to professional and geopolitical impediments.

5

FROM THEORY TO PRACTICE

A standard by definition is:

A document, established by consensus, which provides for common and repeated use, rules, guidelines or characteristics for activities or their results, aimed at the achievement of an acceptable degree of order in a given context.

It is written to be theoretically accurate and completely instructive. However, between a standard and the actual flow measurement, many decisions must be made. It is the application and use of the standard that becomes important. For example, ANSI/API 14.3/AGA Report No.2, Part 2, latest revision provides a table (2-8B) with installation requirements for a "meter tube" (the adjacent upstream and downstream piping attached to the meter). If 29D or more of upstream piping is used as the design basis, the meter meets standard requirements for all beta ratios except 0.75 (Figure 5-1).

However, for a specific meter station with space limitations and known maximum volumes, shorter meter tube lengths may be acceptable. It would be desirable to have the records reflect that such a meter tube was designed for a specific set of circumstances, so that the future users of the tube will be aware of the limitation and not expect it to be a universal length.

In process plants where a tube has defined flow limits that usually will not change, reduced tube lengths are more common than they are in the oil and natural gas industry; an upstream length of 29D is a fairly standard universal length for ANSI/API 14.3/AGA Report No. 3, Part 2, latest revision, although tube length may be even longer in some applications.

The standard's lengths were arrived at by empirical tests, which indicated that shorter lengths caused the coefficient tolerance to exceed the stated limit. Therefore, designs can be made to minimum limits, but the designer should allow some safety factor by using longer lengths. The interpretation that a design is specified by the standards is not true; the standard simply establishes minimum limits. In other words, each

Fluid Flow Measurement. ISBN: 978-0-12-409524-3

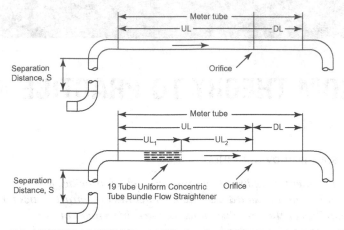

Figure 5-1 ANSI/API 14.3/AGA Report No. 3, a key orifice meter document for flow measurement specialists, provides drawings to help design appropriate meter piping and configurations.

standard is written on the basis of *limits* rather than design specifics.

Those using the standards may not fully understand this subtle difference between "what it says" and "what it means as interpreted by industry design, application, and use." It is worthwhile to seek out these practical interpretations to reach the best design and minimize uncertainty. (Further discussion of *accuracy* versus *uncertainty* will follow in this chapter.)

Ideal Installations

An "ideal" installation is a worthy objective for a meter station. However, seldom, if ever, does such an ideal station exist in the real world. Deviation from ideal starts when meter manufacturers define their meter uncertainties based on the best possible conditions for use—and these conditions are often not reached in actual use, nor are they standardized from one manufacturer to another. The result is that a user should insist on a full and complete disclosure of uncertainty data derivation to make a legitimate comparison between meters from various manufacturers.

Likewise, a user may not take into account all flow peculiarities (such as dirt or pulsation present) not allowed for in the manufacturer's data. It is important for the user to know, and the manufacturer to be informed of, as much of the expected flow application information as possible, in order to derive meaningful uncertainty values for a specific application.

An ideal meter station would be one in which pressure, temperature, and flow are stable—both long and short term—changing less than several percent. The fluid should be clean, of non-changing composition, and with no pulsation. Ample installation space should exist for the required straight meter tube lengths upstream and downstream. Duplicate instrumentation with automatic switchover to standby units in case of primary equipment failure should be included. Instrumentation should include automatic transducer testing, and sufficient periodic maintenance should be planned to reconfirm the meter's uncertainty. Records should be kept to define any outages or anomalies occurring at the station. And meter data should be transferred automatically to the billing or other department with all volumes involved reviewed and accepted by everyone concerned.

Seldom does such an ideal station ever exist in real flow measurement. Therefore, allowances must be made for the non-ideal characteristics and the measurements should be evaluated accordingly.

Non-Ideal Installations

In most cases, requirements for real installations are non-ideal. Very few measuring stations measure an absolutely pure and clean stream of constant composition. Truly clean fluids exist only in designer's minds, so fluid treatment or a meter cleaning system must be provided. Most flowing streams have variable flow rates that must be allowed for in instrumentation selection and measurement system design. And monthly inspection may or may not be frequent enough in dirty flowing conditions. Flow pressure and temperature normally change with time, if not continually. Space is often limited, so required inlet and outlet lengths may have to be compromised. These non-ideal conditions can cause considerable increases in uncertainty, and may well determine design considerations.

Station purchase, installation, and operation/maintenance costs should, but sometimes do not, reflect what the station will be used for—custody transfer, "company use," line/process control, etc. The standards' requirements are not created by considering station uncertainty resulting from noncompliance; they are written so that their requirements are the *minimum* necessary to produce desired measurement uncertainty.

Data on installation requirements are part of the background from which each standard is written. If additional knowledge is desired with respect to a standard's application to a particular design, applicable references should be reviewed. Otherwise, the standards take no position on possible uncertainty values from each design deviation.

Fluid Characteristics Data

In addition to the flow measurement standards issued by various organizations, related data for fluid characteristics can be found in various other references. For example, pressure/volume/temperature (PVT) data are available in the *Manual of Petroleum Measurement Standards*. In addition, the American Society of Testing Materials (ASTM), American Chemical Society (ACS), and National Institute of Standards and Technology (NIST) have pertinent data. Also, universities, as part of advanced-studies programs, have published many correlations of selected fluids.

In each case, the data are based on specific parameter limits of pressure, temperature, and composition ranges. These limits should be known and the data should be used within them, since extrapolations may seriously compromise measurement uncertainty. Equations of these correlations for similar products should also be used carefully; results may not agree because of data uncertainty limits. Each industry uses "accepted data." From time to time, these data are updated on the basis of additional work. The quality and limitations of all such work must be determined before the results are used for good system design.

Limitations of Uncertainty

As previously mentioned, of all the questions brought up at a gathering of flow measurement personnel, the most frequently asked and the least satisfactorily answered is, "What is a meter's accuracy?"

It is an unfortunate fact of life that the one-upmanship often practiced in both the purchase and sale of flow measurement devices may obscure actual meter performance. To adequately define the problem, the following areas of interest must be considered before any discussion is meaningful:

1. Definition of uncertainty;
2. Design of equipment and technical limitations;

3. Manufacturers' adherence to proper techniques when controlling the manufacture of the precision device;
4. Installation of the equipment in such a way as to maintain the manufacturing tolerance;
5. Operation of the station in a manner to produce the best measurement uncertainty;
6. Maintenance required for getting long-term measurement performance;
7. Proper definition of flowing fluid characteristics; and
8. Calculation of volumes from the basic equipment.

Definition of Accuracy

You have seen the term "uncertainty" used in the preceding pages. What is the difference between accuracy and uncertainty? For many decades, accuracy was the term most commonly used to describe a meter's ability to measure flow. It was defined as the ratio of indicated measurement to true measurement. This sounds quite reasonable until an attempt is made to define and demonstrate true flow. Some definitions of "true flow" have included:

1. What the recording chart from an orifice meter says;
2. What the tank gauge says;
3. What the government agency says;
4. What the manufacturer says;
5. What the lab test says; or
6. What I know is right.

An obvious weakness in each definition is the way that it allows a wide variety of answers. Testing by many individuals, manufacturers, research firms, and standards groups has added a large body of information—but not all the results are in agreement. The flow measurement industry had no acceptable statement of exactly how indicated and true values should be obtained or compared. In recent years, a more useful concept has been used: uncertainty. This is defined as a statement of twice the standard deviation of a statistically valid test sample population. This in itself is not an absolute statement of what a given meter will do; it simply states how it will do in some 95% of the cases compared to "the most probable value" as determined by ideal tests.

The test procedure is not specified. The investigator—whether an industry body, a manufacturer, or a governmental agency—sets the test conditions. Results may *appear* "correlated" when fluid is measured once. In industry, however, fluid is normally measured twice: once in and once out of a system. Differences then become apparent.

The second area of caution relates to the accuracy or uncertainty of a meter system compared to a primary measuring device. The user is interested in overall *system* accuracy (i.e., how good is the volume from the system readout), not statements about individual parts—and particularly not a statistical statement based on full-scale accuracy reading of the transducers of a system when measurements are typically not at 100% of range. Without this understanding of the background of the "accuracy numbers game," it is difficult to evaluate statements about a meter's uncertainty made by users and manufacturers.

Most of the numbers that come up in a discussion of flow accuracies are supplied by sources other than the one with the most critical data: *the user under actual field conditions.* The user, then, should be aware of all pertinent factors involved so that a meaningful estimate of likely field measurement accuracy can be made. Properly used flow meters of all types are capable of accuracies that fit in certain categories of proper application. It is the responsibility of those using such meters to fit the meters to actual needs properly.

Design and Technical Limitations

All flow meters use the principles of fluid mechanics to arrive at a value of flow from the fluid's transport properties. Each of these principles has technical as well as practical limitations. For example, the orifice meter is one of a category of meters that requires a pressure drop larger than the pressure drop in normal piping for proper measurement. If insufficient pressure drop is available for measurement, then a head meter cannot be used accurately. This statement seems self-evident; however, users sometimes apply an orifice meter with only a few inches of water differential and still want "accurate" flow measurement. Similarly, an ultrasonic meter that senses velocity must sense an accurate average velocity in relation to a known hydraulic area of the meter opening, or else there is no way to calculate accurate volumes. This means a proper profile must be presented to the meter, and it must be kept clean (Figures 5-2—5-5).

In general, all flow devices are subject to the following limitations, as documented in standards or manufacturers'

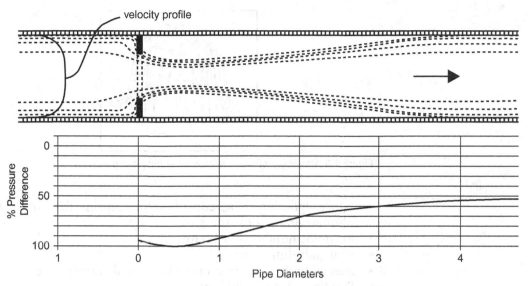

Figure 5-2 Sufficient pressure drop must be created by flowing conditions for valid flow measurement to be derived.

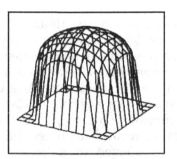

Figure 5-3 Standard fully developed profile.

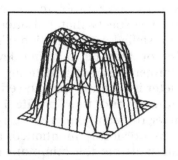

Figure 5-4 Profile following a single elbow.

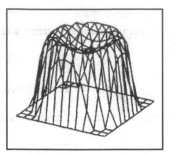

Figure 5-5 Swirling profile—two elbows in different planes.

literature. Ignoring these considerations means that any statement of accuracy is meaningless:
- Reynolds number sensitivity;
- Non-pulsating flow;
- Special piping requirements (flow profile dependency);
- Practical rangeability limits;
- Acceptable calibration data;
- Single phase fluids;
- Accurate measurement of several variables to arrive at standard volume;
- Maintenance requirements with time;
- Recording and/or calculating data correctly.

By coordinating the desired measurement with the above information and relevant standards and/or manufacturer's data, intelligent decisions can be made regarding the possible accuracy that can be expected for a given installation.

The Reynolds number, covered previously, relates how one fluid flow behaves in relation to other fluids with the same number. Meters are affected to a larger or smaller degree depending on the specific meter's response to the flowing Reynolds number. Reynolds number sensitivity should be checked when considering a meter for a given job.

Pulsating flow presents a problem for most flow meters. Whenever a designer suspects that pulsation will be present, the meter must be installed with pulsation eliminators between the source of pulsation and the metering device. Work carried out with most commercially used metering devices indicates that virtually no meter is immune from the effects of pulsation.

Piping requirements and flow profile interrelate. Piping adjacent to the meter run can help create a proper, fully developed flow profile. As previously mentioned, the lengths specified by various standards are the minimum required, and any additional straight pipe will simply add confidence that the measurement is not affected.

Meter manufacturers should—and most do—control all design variables found to affect development of the proper flow profile. However, the care that a manufacturer puts into a meter is useless if proper installation, operation, and maintenance procedures are not followed by the user.

Any installation with meter tube lengths less than those required by standards or manufacturer requirements will result in unpredictable performance and hence should not be used without an in-place calibration.

Measurement with minimum uncertainty will usually occur at the upper range of a meter. Any physical location where measured variables tend to be stable will be better than those where wide fluctuations occur. Measurement will be aided by regulating pressure, stabilizing temperature, and ensuring consistent flow-stream composition.

Once again, standards or the manufacturer's recommended guidelines should be followed for establishing tolerance in the manufacture of meter tubes.

The way that the primary element is attached to the meter tube is important. For example, fabrication should begin with properly selected pipe or meter run tubing. Heat from welding can cause distortion at critical points, so unless proper welding techniques are used, a unit that will assure the minimum uncertainty cannot be produced.

The meter and adjacent piping must be properly aligned. If gaskets are used, they should be undercut by approximately 1/8 inch to prevent an extrusion of the gasket into the line when bolts are tightened. Seemingly insignificant items of this nature cannot be overlooked if minimum uncertainty is desired from the primary measurement device.

As previously noted, recommendations concerning the upstream and downstream piping of a meter tube are covered quite thoroughly in standards and manufacturers' literature. No attempt will be made here to duplicate this coverage, other than to emphasize again that meter tube lengths for all measurement conditions will be best obtained by using the extreme conditions as the design standard minimum (Figure 5-6).

Gauge lines are also important. On differential devices in which the primary element is designed and installed to give an accurate differential at the taps, proper lead lines must also be installed to ensure the inherent minimum uncertainty of the primary device.

Several considerations in the design and installation of these lines for gas measurement must be noted. For gas measurement applications, taps should come off the top of the line, or at least no more than 45 degrees from the vertical of a gas measuring

Figure 5-6 Good design calls for long, straight meter tube lengths for the most accurate flow measurement with minimum uncertainty.

line. The connecting lines should have a diameter of at least 3/8 inch, be as short in length as convenient, and have no direction or diameter changes, in order to minimize leaks and pulsation effects. They should be installed with an upward slope away from the line of at least 1 inch per foot of tubing length, with the differential device located above the line. This facilitates drainage of any condensable fluids back into the line.

The presence of liquid blockage in sections of these gauge lines can cause bias in the order of the equivalent head of water (an inch of water is equal to an inch of differential bias). All gases at flowing temperatures above ambient with pressures near condensation have this problem.

Natural gas saturated with water poses the same problem, even though natural gas itself may not condense. A cold night or a cold rain can cause the entire gauge line and instrument to fill with fluid; then, on warm up, the fluid will evaporate. During this time the instrument indication will be biased in determining rate or total flow. This is more critical for a differential transducer than for a static pressure device; however, the installation suggestions above will minimize problems for both devices.

For liquid applications, the lines should come off the bottom half of the pipe, preferably at 45 degrees from the bottom to prevent solids from filling the line and blocking the differential device. The purpose of the installation is to keep the connecting lines full of liquid even though there occasionally may be gas going down the flow line with the liquid. If a liquid is likely to be heated above its vaporization point by the ambient temperature, then some type of insulation should be installed to control the temperature and maintain the liquid leg in the lead lines.

The flow profile, the pattern or "flow signature" (the combination of velocity profile, swirl, and turbulence) at the meter inlet, is very important for accurate measurement. Two factors control this pattern:

a. The piping configuration—including length, roundness, and smoothness—and the nearest pipe fitting such as elbows, valves, tees.

b. Reynolds number (see Chapter 2)—which is the guide to the shape, size, and stability of the inlet pattern.

Fortunately, most gas is handled at relatively high Reynolds numbers (above 10,000), so that the internal viscous forces are seldom a major component of the predominant inertial forces. A high Reynolds number range is one in which the flow pattern is easily stabilized, provided that the piping is properly installed.

Liquids can have a range of Reynolds numbers depending on their viscosity. A liquid that has a higher viscosity than water should be checked to make sure its Reynolds number is higher than is required for the particular meter. Some meters are specifically designed to operate at high viscosity where the Reynolds numbers are low.

Practical rangeability limits vary with the meter and measurement conditions. A single meter has a limited range for the accurate flow determination, and both the high and low extremes of this should not be approached. It is important to examine a meter's response at very low flow rates carefully. An application's rangeability can be extended by the use of multiple meters in situations where wide variations in flow are experienced and minimum uncertainty is required over the full flow range (Figure 5-7).

A head meter, as the name implies, requires the sacrifice of some pressure, which is of little significance in most fluid measurement, but can be a severe limitation at low operating pressures.

The current standards limit the orifice meter run inside diameter size to 2 inches minimum and 30 inches maximum. Likewise, smaller sizes (as low as 1/4 inch diameter) may be used with special designs, but they are not covered by oil and gas industry standards. Larger sizes are also not covered in standards but are used on the basis of extrapolated data. Turbine meters are available in different limited sizes from various manufacturers, with the largest meter made in the United States presently limited to 12 inches in gas applications and 24 inches in liquid applications. European manufacturers make larger gas

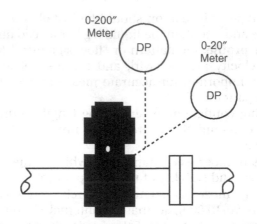

Figure 5-7 If a meter's rangeability is not sufficient to cover the flow range being measured, multiple transducers can be used or a different type of meter could be considered.

turbine meters. Rangeability for turbine meters typically runs from about 10 to 1 on liquid meters and on low-pressure gases. For high-pressure gases, a range to more than 100 to 1 is normal. For positive displacement (PD) meters the rangeability depends on the make of meter, and runs from about 20 to 1 to as much as 1,000 to 1 on both liquids and gases.

If a meter's performance is stable and repeatable, its rangeability may be extended by characterizing its performance electronically.

Standard calibration data depend upon measurements collected over time. Some meters are fully covered by industry standards that have evolved over many years. Calibration data for them have been tested many times. Newer meters have to go through a period of acceptance, and testing normally starts with data supplied by the manufacturer and accepted by the user for non-custody transfer metering.

The supplied coefficients must be carefully evaluated, and checks must made frequently on new meters. Once sufficient experience has been gained and data made available, standards organizations will run their own tests (or accept tests made by others) and prepare a standard.

The industry usually accepts either source of calibration data but will be more cautious about using data from new meters when custody transfer measurement is involved. Standards typically take from five to ten years to complete from the date of development of a meter.

Single phase flow exists in most practical measurement situations. The flow pattern is affected by the presence of two-phase flow, and density is difficult to determine for a non-homogeneous fluid. There are some approximating methods for measuring two-phase flow with head meters, but the resulting data are not precise. The Coriolis meter can measure two-phase flow over limited ranges. There are some new multiphase meters currently being developed to measure three-phase streams (water, oil, gas), but at the time of publication of this book these are just beginning to be used.

Measurement of the other variables needed to derive an accurate standard volume (at "base" conditions) requires attention and understanding equal to that involved with the primary device. The overall accuracy of the flow meter begins with the primary device, but it is also dependent on the transducers necessary to obtain the flowing density either directly with a densitometer or indirectly—through measurement of pressure, temperature, compressibility, and relative density—to convert flowing conditions to base conditions.

Density

As noted, accurate calculation of standard volume through a meter requires knowing the fluid density at the meter and proper interpretation of the measurement through use of appropriate equations to reduce the flow to base conditions (Figure 5-8).

In the past, this calculation of the density from flowing conditions was "standard." Today instruments that can measure density directly are commonly used. A density measurement is needed at flow-sensitive points, such as at the plane of the orifice-plate bore or at the rotor in a turbine meter. A densitometer may be installed in a less sensitive location provided

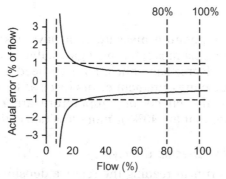

Figure 5-8 Accuracy and other characteristics of transducers are as important as primary meter characteristics for accurate, corrected flow.

Figure 5-9 Metering obtained with improper differential control.

correction or control of the variables is made to arrive at the correct density from the remote location. In cases where a gas chromatograph is used, this can provide density data.

In any event, it is important to always keep in mind that the end product sought is the actual density at the measuring point.

Differential Pressure

Two of the major sources of error in the application of a head meter come from taking the square root of the differential measurement and from the effects of small errors in low differentials, which can cause large errors in flow data. For example, an error of 0.5 inches at 100 inches represents a 0.23% error of flow, at 75 inches it is a 0.33% flow error, but at 10 inches it creates a 2.5% flow error.

Good practice to achieve high accuracy dictates that the differential be kept as high as possible within the strength limitations of the primary device, and the range of flow fluctuation should be limited to the differential measuring device range (Figure 5-9).

Temperature

Errors in temperature measurement have a small effect on head meter flow accuracies; for most gases, there is an 0.1% error in flow rate per degree Fahrenheit. For non-head meters, temperature errors cause measurement errors twice as large as this.

For liquids, the effects of temperature are much smaller, except for those that are 40% or more less dense than water.

Relative Density (Specific Gravity)

An 0.001% error in reading the relative density of a 0.6 natural gas will cause an error of about 0.1% in flow measurement with a head meter; this can introduce fairly large errors in gases with changing compositions, unless the measurement is

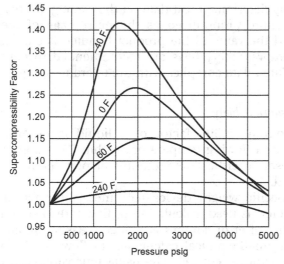

Figure 5-10 Natural gas compressibility and supercompressibility.

integrated into the volume calculation rather than averaged over a time period. This factor also enters into the "accuracy" statement of an orifice meter as a secondary factor in determining the compressibility factor. Relative density only affects the compressibility determination for non-head meters and it does not enter as a direct correction.

For liquids, corrections for the effect of temperature and pressure are related to the measure of relative density or composition and must be used in calculations (Figure 5-10).

Gas Compressibility

The compressibility factor of natural gas (which corrects for the ratio of actual volume to ideal volume) is roughly an 0.5% correction in volume per 100 psi of pressure for an orifice meter under normal pressure and temperature conditions. Hence, an error of several percent in the compressibility factor only produces a small error in volume. However, if the gas is near its critical point, correction factors of *as much as 225%* are required, and small errors in measured variables (temperature and pressure) are reflected as large errors in volume. These values are doubled for non-head meters.

Likewise, gases with large concentrations of non-hydrocarbon gases in their compositions are not as difficult to calculate as accurately, since new data are available from the AGA on these mixtures. Some of the theoretical values obtained by the pseudocritical method (based on the mixture composition) have shown errors of several percent when compared with empirically determined test data on the same gas. This problem becomes more pronounced as the percentage of methane is reduced. If the value of the product handled is sufficient, then actual compressibility tests are recommended to confirm that the calculated data complies with the tolerances required.

Liquid Compressibility

The compressibility factor of a liquid is usually ignored. However, at a flowing temperature within 75% (approximately) of the absolute critical temperature of petroleum, it must be considered.

If a specific weight device (commercially referred to as a densitometer) is used, calculation of flow rate is simplified and the number of error sources is reduced. Assuming an accurate device, the mathematical calculation of flow can be improved and the tolerance reduced. The usual four variables—temperature, pressure, relative density, and compressibility—are reduced to density (if mass is being measured), or density and base relative density (if volume is the desired measurement unit).

In each of the cases cited where there are errors in the measured variables, these arise from two sources:

1. Measurement of the variables.
2. Interpretation of the measurement for conversion to a mass or volume by calculation using the appropriate flow formula (Figure 5-11).

Figure 5-11 Meters and meter tubes must be properly maintained for accurate flow measurement.

Recording and Calculating Data

Recording and calculating data are the final steps in obtaining accurate flow measurements. All secondary devices must be calibrated against a recognized standard. Likewise, when metering devices are exposed to widely varying ambient conditions, calibrations should be made that cover the ranges encountered; if the effects are large enough, consideration should be given to controlling the environment in which the secondary devices operate by adding a housing with cooling and/or heating. The development of new, smart transducers has given the user another option to take care of the problem—but at a higher initial price than for standard transducers. However, the smart transducers are more stable and require less maintenance. A balance between the accuracy required and the cost of obtaining it will determine the extent to which you can justify testing and purchasing smart transducers versus installing housing (Figure 5-12).

Measured data must be either recorded or transferred to a central calculation office for flow rate conversion, or this can be calculated directly by computer equipment installed at either location. Each step of recording or transducing and interpreting adds potential sources of error to the flow measurement. A simpler system *with proper maintenance* is usually found to yield the best results and optimum return on investment.

One of the most serious problems with the use of new types of recording and calculation equipment is the failure of manufacturers and users to recognize that proper personnel training is needed to get accurate measurements with the equipment. Anyone who buys or sells without extensive training in the use of equipment that is "different" or "strange" can be assured that, at worst, the equipment will turn out to be "no good," and at best, that there will be a time period before enough familiarity is gained to allow the equipment's capabilities to be realized.

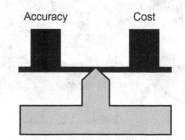

Figure 5-12 The cost of accuracy should be compared with the accuracy needed for the application.

Summary

By following the guidelines in this book and the provisions of the standards, along with recommendations from manufacturers and users regarding similar applications, primary elements can be selected that will offer the best possible accuracy in any specific measurement installation.

Any overview of accurate flow measurement should contain a discussion about what kind of results can be obtained if all precautions are taken. Without full qualification of the data source and complete definition, accuracy statements are meaningless. Proof of accuracy usually comes down to a study of system balances of measured flow inputs versus flow outputs. Experience shows clearly that the only way that acceptable balances are obtained is by following all of the **best** (not good) practices in design, application, installation, operation, maintenance, and interpretation (Figure 5-13).

There is no such thing as absolutely accurate *flow measurement.*

Measurement is always performed to some limit of accuracy. The purpose of any flow measurement should be to measure as accurately as possible within pertinent economic constraints. For a "flow purist," measurement should be completely accurate; but for a commercial flow-system designer, the cost relative to measurement accuracy and system maintenance must be considered. Since flow measurement has a tolerance within budget limits, the designer's function is to minimize this tolerance so that the investment for better accuracy can be demonstrated as desirable.

Figure 5-13 All new equipment must be thoroughly understood if it is to provide maximum effectiveness to the user. Classes are available from manufacturers, measurement schools, and individual operating companies.

Figure 5-14 A well designed metering station (properly installed, operated, processed, and maintained) represents good value for money, whose reliable measurement accuracy will maximize the return on investment.

It is very important that the accuracy is reaffirmable during flowing conditions, so that full confidence in the flow metering system can be enjoyed over extended time periods. An equally important consideration is that the accuracy is actually reaffirmed on a scheduled basis.

A final note should be added on this balance between accuracy and system cost. The law of diminishing returns for flow measurement says that beyond some point, additional expenditure will not buy any provable better measurement (Figure 5-14).

References

American Chemical Society (ACS) Journal of Chemical & Engineering Data, 1155 Sixteenth Street N.W. Washington, DC 20036.

American Petroleum Institute, API Manual of Petroleum Measurement Standards. Publication and Distribution, 1220L St. NW, Washington, DC 20005.

ANSI/API 14.3/AGA Report No. 3, Orifice Metering of Natural Gas and Other Related Hydrocarbon Fluids, Arlington, VA.

ASTM ordering address, 1916. Race St., Philadelphia, PA 19103.

National Institute of Standards & Technology, 325 Broadway, Boulder CO 80303—or Gaithersburg, MD 20899.

FLUIDS

Fluids—Liquids and Gases

If the fluid is water at ambient conditions, then its influence on flow measurement can be easily calculated from known and accepted data. However, if it is a mixture near its critical temperature and critical pressure, then accepted data may not be available, and the volume change with minor changes in temperature and pressure may make fluid definition the most important consideration in obtaining accurate flow measurements. *This is one of the considerations most often overlooked in selecting and using a meter.*

A meter's commercially advertised accuracy normally allows for no error in determining the influence of a fluid's physical properties, and users are misled into believing that simply buying a potentially accurate meter will take care of all problems. If the fluid is not adequately conditioned for flow measurement, then no meter will achieve its potential accuracy. It is of value, therefore, to review important fluid characteristics in order to know how to design an optimum metering system.

Low Risk Flow Measurement Fluids

Low risk fluids:

1. Are not near their flash point (for liquids) or condensing points (for gases);
2. Are clean fluids without other phases present, with a composition whose PVT (pressure/volume/temperature) relationships are well documented with industry-acceptable data;
3. Are not exceptionally hot or cold, since temperature may limit the ability to use certain meters;
4. Have minimal corrosive, erosive, or depositing characteristics.

Considering these characteristics and answering related concerns minimizes the influence of fluids on measurement accuracy, and simplifies metering.

Fluid Flow Measurement. ISBN: 978-0-12-409524-3

High Risk Flow Measurement Fluids

Many fluids classified as "high risk to measure" become the main consideration in the choice of a meter and in determining the potential for measurement accuracy.

High risk fluids include:

1. Two or more phases in the flow stream;
2. Dirty mixtures;
3. Flows near fluid critical points;
4. Flows with temperatures over 120° or under 32°F;
5. Highly corrosive or erosive fluids;
6. Highly disturbed flows;
7. Pulsating flows;
8. Flows that undergo chemical or mechanical changes; and
9. Highly viscous flows.

Specific meters may react differently to the problems listed above, and there may be one that works better than others for the specific problem presented. It should be recognized that the fluid sometimes must be measured even if it is a "high risk" fluid, and the cost of making it a "low risk" fluid is prohibitive in a cost/value study. The preferred fluid conditions are sometimes simply not available at the point of measurement (Figure 6-1). On the other hand, these characteristics may result in a measurement uncertainty that is no better than ± 20 to ± 30%. An example is the measurement of carbon dioxide for injection into oil reservoirs for tertiary oil recovery. Removing oil from the formation efficiently requires the CO_2 to be in the

Figure 6-1 The water that comes out of your tap at home is a "low risk" fluid to measure.

"dense phase" stage (near the critical point). Under these conditions—where the temperature and pressure are near the critical points of 88°F and 1,087 psia—the CO_2 density variation may be in the order of one percent per degree Fahrenheit. In this high risk fluid situation, the flow measurement practitioner might prefer to change the temperature, or the pressure, or both, but the successful use of the fluid to remove the oil precludes such changes, so wider limits must be put on this measurement.

Basic Requirements and Assumptions

Fluid flow must be single phase for basic meter accuracy to be meaningful. There are two problems with a two-phase fluid. One is its effect on the meter mechanics, and the other is difficulty of obtaining a truly representative sample to determine the composition for calculating the reduction to base conditions. Studies are under way to determine whether there is a recommended way to sample two-phase fluids. Currently, sampling under these fluid conditions is less than optimal. This is particularly important in oil and gas production operations (Figure 6-2).

For liquids in a gas, the two-phase flow may affect flow profile configurations. Such flow patterns may conform to one of a number of regimes in a pipeline. The first phase tends to be droplets; if they are small enough, they form a homogeneous mixture so both the flow effects and sampling errors for composition mentioned above are minimized and may not be significant. The second regime occurs when there are sufficient droplets for them to begin to accumulate in non-flowing areas, including in the area of the meter, where flow distortion may occur. Also, a sample can be inaccurate. The third regime occurs when additional liquid is present. For example, there may be annular liquid flow with a core flow of gas. In this case, the mechanical configuration and the sampling are in trouble.

The addition of even more liquid means that there will then be two separate flowing streams (usually flowing at different velocities) forming layered flow (Figure 6-3).

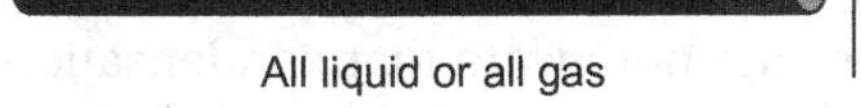

All liquid or all gas

Figure 6-2 Single phase flow is required.

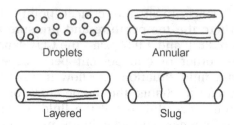

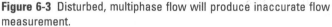

Figure 6-3 Disturbed, multiphase flow will produce inaccurate flow measurement.

The next regime is slug flow, which occurs as liquids collect until the lines are filled to a point where they "burp" the liquids. In such cases, there will be no way to take care of mechanical or sampling problems since liquid slugs will be followed by gaseous pockets with no indication of which fluid is causing the meter to indicate flow. Separation must be done prior to attempting measurement in any of the cases where two-phase flow exists at the meter.

In the reverse case—gas in liquids—similar problems exist, and de-aerators must be used. With solids in liquids, filtering is required to correct measurement and to minimize meter damage.

Some meters that react to mass can be successfully used for two-phase measurement provided they are not used to attempt to calculate volume without additional information on the fluid composition or density.

Some fluids are unstable and may present measurement problems when they break down into other products during interrupted flow conditions, or during exposure to conditions along a line. Hydrates in natural gas measurement are an example. Hydrates, a mechanical combination of hydrocarbons and water that forms an ice-like material, can block off main lines or gauge lines so that readings of differential or static pressure are impossible. Likewise, crude oil can form an emulsion with water, which does not lend itself to flow measurement. Some liquid plastics and such fluids as molten sulfur set up if they are not kept flowing or the meters are not heated. In these cases, the fluids must be removed from the lines when shutdown occurs. The mechanical problems these fluids create may prevent measurement if these precautions are not taken.

Although the problems caused by two phases have been outlined, some users are not aware that condensation of gases often may take place in a meter, since this may be the point with the lowest pressure and temperature in the system. Although the

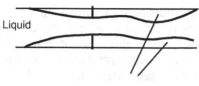

Figure 6-4 Flashing fluids cannot be measured accurately.

fluid may be above saturation upstream in the system, conditions at the meter may be below the condensing point.

The same is true with flashing liquids; the pressure within the meter may be lower than the pressure existing after flow goes through the meter to "recovery." If there is any question, a higher back pressure should be planned, rather than risking the fluid flashing (Figure 6-4).

Fluids that are measured near their critical point have variations in relationships which are so large that the measurement of pressures and temperature within tolerances close enough to predict the effect on volume is beyond the ability of measuring equipment. Likewise, the accuracy of PVT correlation data deteriorates as the critical pressures and temperatures are approached. Because of these problems, flow measurement should not be attempted in such areas of operation if this can be avoided. The problem may also be addressed by moving the meter station to a better location in terms of fluid properties.

A number of meters have limitations for lower Reynolds numbers, and such limits should be checked in the standards prior to attempting to use a meter. With new meters the information on these limits is often quite general, and the user may be left in a quandary as to its meaning. If there is any question, a meter should be operated well away from such limits for best accuracy.

In summary: from a measurement standpoint, no fluid should be measured near a point of phase change, fluid characteristics change, near condensation, at too low a Reynolds number, or near critical pressure or temperature. Good flow measurement practice requires that these conditions are recognized and proper precautions are taken to minimize their effects.

Data Sources

Many sources of data for fluid physical properties characteristics are available in the industry. The particular industry standards for specific applications should be the first place to look.

There are general fluid specifications available from universities, national standards organizations, and from suppliers that handle certain products; manufacturers of fluid products have their own data. The *Flow Measurement Engineering Handbook* has accumulated much data that is useful for general flow measurement.

Gas and liquids are usually considered as either "pure products" or "commercial products" in most references. Mixture laws may be used to estimate the combined characteristics of uncommon fluids, but caution should be exercised if the mixtures contain widely varying molecular weights or involve extreme conditions of pressure and/or temperature. In these cases, actual PVT tests should be run over the ranges of operation and a specific set of tables or equations should be set up for the particular mix—if such information can be derived from the results obtained.

It is important to avoid the tendency in industry to specify a mixed fluid only by its major component. Mixture characteristics must be considered, not just the PVT relationships of the pure product. Any data correlation must be examined to determine the fluid data parameters involved before applying the data to a specific metering system (Figure 6-5).

Liquid propane is an example of a liquid that falls into this area of concern. Commercial propane contains 95% propane, while the percentages of the other constituents may vary; this variation will affect the correction factors used. Reagent grade propane, however, allows the use of "pure" correction factors such as those available from the National Institute of Standards and Technology (NIST).

Figure 6-5 With marbles representing fluid molecules, it is easy to see that mixing an equal volume of two different fluids does not necessarily result in twice the volume. This phenomenon is more prevalent in liquids and in some applications is known as shrinkage.

Another way to approach the problem of variable fluid characteristics is to use densitometers or mass meters, and then measure mass. The measured mass can then be converted to an equivalent volume at base conditions using an analysis and the density of the individual components at base conditions. (Note: This approach will be correct to determine a contract volume. But if there are significant differences in the flowing and base volumes, some means of deriving an approximate line condition flow volume may be required for operations.) The contract volume is necessary for changing custody, but the *actual* volume at line conditions is needed for operational control.

For any petroleum fluid, Chapter 11 of the API *Manual of Petroleum Measurement Standards* is an excellent data source.

The ASTM, in conjunction with both the API and the GPA, also publishes common standards for petroleum-related fluids. International standards organizations also publish fluid data documents. (See Chapter 4, "Basic Reference Standards" in addition to the list below.)

Fluid Characteristics

Gases

The following section discusses problems unique to some commonly measured gases (Figure 6-6).

Natural gas is one of the gases most commonly measured, since it is used both as a fuel and a feedstock in many industries. It represents the largest daily dollar volume of any gas routinely bought and sold. The natural gas industry has been

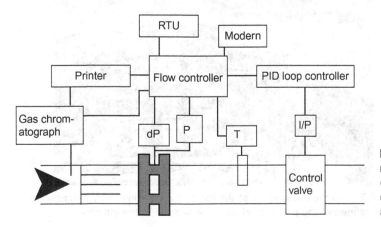

Figure 6-6 Modern natural gas measuring systems may be quite complex, and all elements must be considered in determining metering accuracy.

the leader in developing gas measurement technology and its standards are used in related areas of gas measurement.

The ANSI/API *Manual of Petroleum Measurement Standards*, Chapter 14, "Natural Gas Fluids Measurement," Section 3, "Concentric, Square Edged Orifice Meters" Parts 1–4 (also entitled "American Gas Association Report Number 3," Parts 1–4; and "Gas Processors Association 8185" Parts 1–4) is the most common standard used for gas measurement of all kinds. Representing over 70 years of study of gas measurement with the orifice meter, the standard continues to be upgraded further by additional work.

Natural gas as a fluid varies from an easily measured fluid to a very difficult fluid to measure to close limits. Separated, dehydrated, "pipeline quality" natural gas is normally easy to measure since it is a well defined fluid for which very precise data is available for relating the pressure, volume, and temperature from flowing conditions to base conditions. It is normally very clean, with a minimum of solid "pipeline dust" and compressor or dehydration plant oils present. It is normally handled at temperatures and pressures that cause minimal meter design limitations or operating concerns.

Pipelines typically measure the gases within tenths of a percent in their pipeline balances between purchased gas, operating use gas, and gas sold (Figure 6-7).

On the other hand, produced gas is often handled as a saturated fluid (separated to single phase but not dried), and the

Figure 6-7 Flow measurement is at the heart of controlling many processes in modern refineries and chemical plants as well as for other industries.

problems of flow measurement increase. The ability to balance a production field (with multiple wells) is thought to be within the industry norm if this balance is within 3 to 5%.

Other than some operating problems and poor maintenance that may affect measurement, the main cause of error is the fluid characteristics that cause both mechanical problems (liquid in the meter) and errors in fluid density calculation (determining the proper specific gravity or relative density from a sample). Quite often, the volumes measured are not used directly for custody transfer, but for allocation in determining the percent of total volume contributed by each well's flow.

At the present time, the measurement of two-phase fluid (gas and gas liquids and/or water) is not attempted because of the problems caused in a meter. There is a great deal of work being done on multiphase measurement that may, in time, resolve these problems. Getting a material balance in a system or processing plant can be one of the most frustrating flow measurement jobs. Careful attention must be paid to all of the concerns outlined here; even then, getting a close balance is difficult.

Mixtures of gases are more easily measured if the mixture has a relatively constant composition. This allows specific PVT tests to be run, or data may be available for common mixtures from prior work. The ability to use the mixture laws successfully has been previously discussed. If the mixture is changing rapidly, use of a densitometer or a mass meter may be required to determine an accurate quantity.

Ethane, a common chemical building block, may be a measured product or a mixture of an enriched stream from a processing plant. Data are available on ethane as a pure gas product. Except at critical points, measuring ethane is fairly straightforward. However, ethane's critical temperature of 90.1°F at a critical pressure of 667.8 psia may easily occur in normal pipeline and process plant operations. For the best measurement, some heating or compression may be required for conditions near critical to make them more favorable for measurement.

From a measurement standpoint, it is worth remembering that though classified as a gas, ethane mixtures have the same characteristics as dense fluids up to approximately 1,000 psia and at temperatures from 90° to 120°F. In this area of operation (which may exist quite often in a pipeline, or as part of a pipeline or processing measurement requirement), the density changes significantly with small changes of temperature and/or pressure. Because of this, flow metering accuracy relates to these and proper selection of the equation of state for the ethane. At pressures below 670 psia, if the temperature drops below 90°F, two-phase flow can be encountered. When either of these situations is likely, consideration should be given to adding heat or pressure to the system prior to attempting flow measurement.

Measurement of ethane as a pure liquid is not very common. Since the liquid has to be handled at a low temperature, it presents unique problems with meters. However, the data for making PVT corrections are available and accepted.

The measurement of ethane liquid mixtures is more common. As noted above, data on pure ethane are readily available, whereas data on ethane-rich streams are limited, and metering accuracy will suffer accordingly. This is particularly true for variable mixtures. At times the better option is to use a densitometer with careful attention to proper sampling to minimize sample errors due to temperature and pressure variations in the stream.

Propane can be handled as a liquid or a gas since its critical temperature is 206°F at a pressure of 616 psia. At normal ambient temperature, it can be a gas or liquid, depending on the pressure. However, to solve measurement problems, the phase relationship must be known so that a single phase of liquid or gas is considered. The meter can then be properly sized, and the two-phase or phase-changing regions can be avoided. As with any fluid, the closer to phase change the measurement is attempted, the more difficult measurement becomes.

Ethylene is a popular hydrocarbon feed stock used in the chemical industry. It is difficult to measure in a number of industrial cases since its critical temperature is 48.5°F at 731 psia, which means that it is handled in an area of greatest density sensitivity (i.e., the measurement problem becomes one of the correct density measurement). Near this point in the vapor phase, the compressibility factor changes as rapidly as 1% per degree Fahrenheit and 0.5% per 1 psi pressure. This problem is significant enough to, at times, require heating prior to flow measurement to obtain sufficient accuracy. Heating to 80° to 90°F will minimize these density changes. This represents an example where the effects of the fluid characteristics on measurement accuracy are so significant that they are changed prior to measurement.

Another characteristic of ethylene is that a very fine carbon dust is produced during processing, which cannot be removed with a 5 micron filter. Depending on the frequency of buildup, the meter and meter piping must be cleaned so that the flow characteristics are not changed. This also affects measurement transducers and their lead lines, which must be cleaned.

Ethylene often contains small quantities of hydrogen. This can affect filled differential transducers; an internal pressure builds up over a period of time to the point that the unit will not be capable of meeting calibration, and replacement is required. How often this happens varies with pressure and hydrogen content.

Ethylene requires that special elastomer materials are used for those parts of a meter that come in contact with the fluid ("wetted" parts). Meter materials should be checked, and meters should be ordered to accommodate these requirements.

Propylene is another popular feed stock for the chemical industry. It is somewhat easier to measure than ethylene. This is true because its critical temperature is 197°F at a critical pressure of 667 psia. Propylene is less reactive with most meter materials, but materials in meter seals should be checked carefully for reaction with propylene.

Carbon dioxide, commonly measured, is used in the oil and gas industry for the recovery of crude oil. It also has a troublesome critical temperature of 88°F at a critical pressure of 1,071 psia that is significant when attempting flow measurement. The compressibility factors in these ranges may represent as much as a 200% correction and will be the controlling factor in achieving accurate flow measurement. Carbon dioxide is quite often handled as a mixture, which further complicates the measurement of density. Data are available from the NIST in Boulder, Colorado, for mixtures of CO_2 from 94 to 99.7% containing small amounts of methane (0 to 2%), ethane (0 to 1%), propane (0 to 2%), and nitrogen (0 to 2%), as well as pure CO_2 (Figure 6-8).

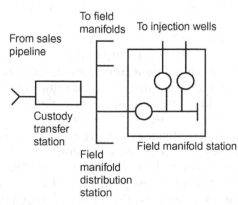

Figure 6-8 Typical CO_2 injection system for crude oil recovery.

Once again, carbon dioxide measurement is not easy because of the density sensitivity at normal operating conditions—even if well removed from the two-phase flow region. The solution requires continual integration of density with the flow device because of the rapid changes. A computer is needed to do this calculation rapidly enough.

Another measurement problem is CO_2 wetness. If water is present at a sufficient level, a hydrate may be formed at temperatures well above freezing (32°F). In addition, wet CO_2 is very corrosive, and a large amount of corrosion products will move with the gas; this often results in deposits that can cause problems with meters and other equipment such as densitometers. Likewise, CO_2 causes most standard seal materials, such as rubber and Teflon, to break down, and lubricants rapidly deteriorate in the presence of CO_2. All these factors should be taken into account before designing a carbon dioxide metering station.

Steam flow measurement is often one of the most misunderstood of all flow measurements made in industry. There are many reasons for this, but fluid problems are the most important. Measuring steam as a fluid is fundamentally the same as measuring any flowing fluid. If the fluid dynamics and the thermodynamics are known, then the first steps toward accurate measurement have been taken. However, these two areas are *not* particularly well known or understood in many of the applications in which steam measurement is required.

As commonly used, the term "steam" is meaningless when considering measurement. In the industrial world, the definition narrows somewhat, but even here it covers a variety of flowing conditions. The following sections cover the three possible steam flowing conditions: wet, saturated, and superheated.

Wet (quality) steam is usually the most difficult fluid to measure. A wet steam is a fluid that contains both condensed hot water and steam. In the two-phase portion of the phase diagram (see Figure 6-9), with the same temperature and pressure there is a different density. Therefore a third parameter, quality, must be added to correct a measurement for the right density. Quality is defined as the ratio of percent flow that is steam to the percent flow that is free water, in mass. For example, 95% quality means that 95% of the flow is steam and 5% is water by mass.

Quality can be determined by a calorimeter test (batch operation), which is valid only until changes in the system take place (i.e., flow rate or density change). In addition to these fluid identification problems, the effect of two phases on the meter mechanism can create errors. Because of these problems,

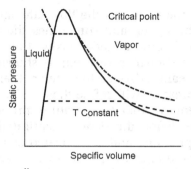

Figure 6-9 Steam phase diagram.

Table 6-1

	Temperature	Pressure	Quality
Wet	Yes	Yes	Yes
Saturated	T or P	T or P	—
Superheated	Yes	Yes	—

quality steam measurement is inaccurate and should be attempted only as a last resort, recognizing that it will have very wide accuracy tolerances.

Saturated steam has no free water present and exists only at one pressure and a corresponding temperature. However, water is not always pure; therefore the exact saturation point, pressure and temperature may be difficult to define. At the same pressure and a higher temperature, the steam is *superheated;* at a lower temperature, condensation takes place, and the fluid becomes quality steam. Saturated steam exists at a boiler, but when steam leaves the boiler (assuming no superheat is added), the flow creates a pressure drop, and there is the possibility of a temperature drop, depending on flow line insulation. Therefore, steam traveling through a plant will normally be superheated (i.e., the pressure drops, but the temperature is relatively constant and the steam is almost never saturated away from the boiler). Table 6-1 shows the measurements required to determine steam density.

Most designers state that they will be handling "saturated steam" and may not allow for temperature and pressure measurement at the meter. As the flow rate varies, the pressure will change at the meter, and sometimes the temperature will also change. To determine the density, the temperature and pressure must be known (measured) as shown in Table 6-1. If true saturated steam exists at the meter, then measuring the temperature or pressure will define the density.

Wet steam requires that temperature, pressure, and quality are measured to define the density. Saturated steam requires temperature or pressure, and superheated steam requires temperature and pressure. In each case, these measurements must be fed to a computer that calculates the density based on the equations in steam tables.

A problem unique to steam is the large difference between the temperatures of ambient air surrounding a meter and the flowing fluid. This makes the proper measurement of steam temperature a major concern. Without suitable insulation and special precautions, major errors in density will result. In a flowing stream at low velocity, steam tends to stratify by temperature; the steam must be mixed to get the temperature constant across the stream and allow accurate steam measurement (Table 6-1).

In summary, steam is the most difficult fluid to measure accurately. Even with the utmost care, a plant balance of steam and water flows in a power-generating or process plant is very difficult to do well.

Liquids

Liquids are generally reputed to be easier to measure than gases. That might be true if all liquids behaved like water at ambient temperatures. They do not, however, and certain precautions need to be taken for optimal flow measurement of each of the fluids that will be discussed in this section.

Crude oil has had as much research conducted on it as any liquid in terms of product value as well as related to the worldwide measurement and product exchange. The generic term "crude oil" covers a multitude of fluids that can be categorized by the following terms: light, sweet, sour, and waxy. These conditions all affect flow measurement procedures.

Light crude is the most desirable from a measurement point of view. Its viscosity range and wax content are the lowest, and both of these affect metering. Some heavy crudes cannot be measured without heating. The high wax content creates the possibility of deposits in lines and meters, and no common meter can operate properly if this occurs. Wax treatment or heating is required before measurement is attempted. As previously outlined, flowing temperatures must be known to define the magnitude of measurement problems. A statement about light crude must be accompanied by the operating temperature range to allow proper metering system design.

Sweet and sour crude oils typically affect meter materials and involve foreign materials introduced by corrosion. Knowledge of the fluid composition allows meter materials to be selected appropriately. Corrosive products may be treated or filtered out before they enter the meter.

Most meters are sensitive to viscosity, which limits the range some can handle. The effect of temperature on viscosity adds to the problem—plus the fact that the temperatures of crude oil measurement cover a wide range, and they are normally not controlled and are therefore determined by the situation. For example, storage tank oil may run at 120°F, whereas tanker oil will arrive at ocean temperature, and pipeline flows typically arrive at ground or ambient temperature.

Medium and heavy crudes have intermediate characteristics, but high viscosities and "crud" content can aggravate metering problems.

Complete pressure and temperature correction data are available in the literature and have been accepted by petroleum measurement groups across the world. These data were first accepted in early 1980 (in the United States, August 1980) and are referenced by most contracts. The data allow corrections from flowing conditions to base conditions not only for crude oil but also for all liquid hydrocarbons within the defined data base limits.

Dirty crude includes foreign material that will collect in a meter. This is often experienced in some production areas where the term "grass" is applied to the phenomenon. Such materials should be filtered out before metering to prevent meter stoppage or inaccuracy. If "pipeline quality" fluid is achieved, this problem seldom occurs.

Refined products, as the name indicates, are processed so that foreign materials have been eliminated. There are normally limited amounts of various components, since refined products must meet industry-established product specifications. A large quantity of industry data is available (as mentioned in the crude oil section). For new products, individual buyers and sellers will develop accepted data based on PVT tests covering the ranges of operation. Then, when the product becomes widely traded, the industry will correlate the various data, run additional tests as necessary, and establish industry correlations. Hence, most of these fluids have well-defined characteristics that can be used with confidence to get good measurement.

Ethylene and propylene liquids are well defined if they are pure products. Unless they are handled near their critical temperatures, they introduce no specific problems. Concerns about measurement near the critical points of these two fluids, or for their mixture problems when they are not pure, have been pointed out previously in the section relating to the individual compounds.

Gasoline is relatively easy to measure since it is stable over the temperature ranges at which is it normally handled, and has no tendency to flash to a gas. Correction factors are well established and accepted, and they are available for many meter readout systems with no special programming required. There are no viscosity or foreign material problems to cause special concern about a meter's operation.

Heavier hydrocarbons (i.e., C_{10} and heavier) present potential viscosity problems that must be addressed, since they are normally handled in smaller quantities, and small meters have more problems with viscosity. These fluids also may be unstable at normal handling temperatures and require controlled temperature and pressure conditions.

Natural gas liquid mixtures produced from natural gas contain variable amounts of light hydrocarbons (ethane through nonane). They are significant fluids in the oil and gas industry and have received a great deal of attention for flow measurement. As indicated in their definition, natural gas liquids are undefined mixtures that quite often change composition over time, pressure, and temperature. They are recovered from separators whose efficiencies relate to operating temperatures and pressures that change between night and day. The range of variations in flowing temperatures and pressures can be wide, which further complicates measurement. With the wide range of molecular weights of some of the components—particularly ethane (30) and pentane (72)—mixtures of the products have variable shrinkages with changing compositions (see Figure 6-5).

The problems above have given rise to a number of individual operating company pressure and temperature correction tables as well as tables available from the Gas Processors Association that apply to a restricted data set. Since these problems arise from changes in operating parameters, many operators have opted to use a densitometer with a volume flow meter to measure mass flow, or a mass meter that measures mass rate. A true mass meter does not require a densitometer. If an

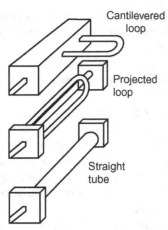

Cantilevered
loop

Projected
loop

Straight
tube

Figure 6-10 Types of Coriolis true direct mass flow meters.

analysis is made, the mass flow can be converted to volume flow by knowing the cubic feet per pound at base conditions of the individual components (Figure 6-10).

The amount of mass or volume to be measured will define the equipment required. With wide variations in fluid characteristics, the procedure of using mass and analysis provides the most accurate way of measuring these flows, particularly at extreme temperatures or pressures.

When natural gas liquids are processed, they can be broken down into their pure components, which are more fully defined liquids. However, it is very important that quality requirements for the products be taken into consideration since they will affect the flow measurement. If the fluid is reagent grade ethane, propane, or butane, then industry-accepted pure product corrections may be used for correcting from flowing to base conditions. On the other hand, if commercial grade product mixtures of hydrocarbons are involved, then mixtures of hydrocarbons exist, and the correlations of the pure products must be adjusted to reflect corrections from the pure products' specific gravities. Here again, industry-accepted correction tables may be used within the limits of their databases.

When a customer states he wants to measure "propane," the meter designer does not know exactly what is required. Reagent grade is over 99.5% propane, but commercial propane requires only 95% propane. The amount of ethane and butane may vary and cause correction factors to vary. So-called propane-rich streams may have even *lower* propane percentages.

Ethane-propane mixtures present a measurement problem with fluid PVT relationships, since normal handling conditions are near critical conditions.

There are numerous correction tables from operating companies based on their own databases. Some of these are available from the Gas Processors Association. There is also a study taking place to standardize the temperature correction values in liquid. The specific gravity range of 0.35 to 0.70, with a temperature range of roughly 50° to 150°F, generally covers the EP mix. The purpose of this study is to compare all standard procedures and, if possible, pull the industry together to agree on a single relationship. (Availability was imminent when this book was being published, in 2014.)

Two-phase flows fall into two general categories with most meters: measurable and not measurable. Since current techniques do not always provide the ability to prevent two-phase flow, studies have been made of handling the problem in limited ranges. Within these specified limits, the methods have been correlated based on the density of the two individual streams, so as to address the problem of up to 5% by volume of gas in liquids and up to 2% by weight of liquids in gas. These are very low limits and should not be stretched to higher-content mixed flow. In each of these cases, the accuracy tolerance of such measurement is at least double that expected by single phase measurement with a given meter. These procedures have been applied to steam and condensed water systems, natural gas and natural gas liquids, and crude oil and gas flows.

Some true direct mass meters can measure two-phase flows within design limits. Flows outside these limits are not measurable and should be separated and measured as individual liquid and gas flows.

References

American Petroleum Institute. API Manual of Petroleum Measurement Standards, Chapter 11, Physical Properties Data Chapter 14, Natural Gas Measurement, Section 3, Concentric, Square-edged Orifice Meters, Parts 1−4.

American Society for Testing and Materials. 1916 Race St., Philadelphia, PA 18103.

ASME. Steam Tables In: Thermodynamic and Transport Properties of Steam. New York, NY.

Benedict, Webb, and Rubin. 1940. Equation of State In: Journal of Chemistry & Physics. 8, 334.

Fischer and Porter. 1953. Catalog 10-A-94. Warnister, Pa. (Contains liquid specific gravities and viscosities.)

7

FLOW

The flow characteristics of a fluid can aid or detract from the ease of making a measurement with a flow meter. Certain basic assumptions made previously in this book are amplified below.

Required Characteristics

The required characteristics of the flow include: continuous, non-fluctuating, non-pulsating, and the pipe running full in liquid flow.

"Continuous" means that the flow should not continually stop and start. Each meter has a certain amount of inertia to start, plus overshoot after a flow stops. Furthermore, during these periods, the inaccuracies of the low rate measurements are greater than the values typically quoted by manufacturers for specific meters.

On the other hand, startup and stop requirements for flow measurement are common for batch-type operations. No meter measures correctly from zero flow to normal flow. However, when totalized flow is the desired goal, short startup and shutdown times can be insignificant to a total flow, provided they correspond to a small percentage of the total flow time. Tanker loading of crude oil is an example of this type of operation (Figure 7-1).

For example, loading a supertanker may take 12 hours, and a start of eight minutes may be needed to get up to loading rate. It may take less than two minutes of the eight before a meter gets into its accurate range. This represents less than a few tenths of the total loading time and less than a hundredth of the total fluid loaded. Such a small error is usually compatible with the inaccuracy of the measuring system and typically can be ignored.

On the other hand, if a large portion of the metering time and/or much of the total flow happens at low rates, then a multiple meter system should be designed, or alternate meters should be evaluated. For large swings, the system might have

Fluid Flow Measurement. ISBN: 978-0-12-409524-3

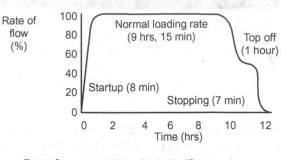

Typical flow rates for tanker loading

Errors from startup/stopping insignificant
because of low flow and short duration

Figure 7-1 Short periods of inaccurate flow measurement can be unimportant compared to very large total volumes.

one normal meter, one meter to measure at the low rate, and a third to handle an occasional peak. The key design parameter should be to minimize the percentage of the total measured flow represented by excursions of the flows.

When on-off flow operating control systems are required, then extended-flow-range metering systems must be used.

Fluctuating flow presents a problem when its rate of change falls outside the response time of the metering system. In such a case, the meter may *appear* to respond correctly, but the measurement accuracy can be severely compromised. Dampening readout systems to improve flow metering can severely increase measurement uncertainty. So much dampening can be applied that any meaningful indication of the actual flow is obliterated.

The proper method of dealing with this problem is to dampen the stream-flow variation so that the fluctuations remain within the meter system's response time. If this is not possible, then a faster metering response time will be required.

Pulsation versus fluctuation relates to the frequency of flow changes. Pulsations may also come from pressure changes not directly related to flow but which can cause metering errors. In either case, the complexity of the problem strongly suggests that these variations in flow and/or pressure be removed before flow measurement is attempted. This problem is more prevalent in gas than liquid measurement.

Most commercially available metering systems do not have a response time fast enough to respond to flow pulsations more rapid than a few hertz. The effects of pump, compressor, control valve, or piping-created pulsations may exceed the meter's response time. Here again, improper dampening of the readout system may make an operator happy with the flow record, but can introduce major flow readout errors—of *over 100%*.

Since recognizing pulsations significant enough to cause error in flow is very difficult, several instruments have been developed to help predict whether or not pulsation is causing flow measurement problems. None of these is capable of being used as a correction device; they are used only to discover pulsation and/or to show that the pulsation has been eliminated sufficiently to allow valid flow measurement.

Designing piping properly so as to minimize acoustic "tuning" in different meter systems can be a useful approach to minimizing problems in both the primary meter piping as well as in the secondary instrumentation. Experience has shown that a majority of the flow measurement problems with head meters caused by pulsation occur in the secondary instruments. Short, large-diameter piping to the differential meter is recommended, since such piping tunes only to very high frequencies (over 100 hertz), which are normally above the frequency that causes large flow errors. When pulsation is present, all statements of meter system accuracy are suspect. The pulsations should be eliminated before flow measurement is attempted (Figure 7-2).

Full-conduit flow is important in liquid systems. The flowing pipe must run full, or any measurements made will be in error. This can be a problem if piping design does not keep the meter below the rest of the piping. If the meter is at the high point, then vapor can collect and create a void in the meter so any velocity or volume displacement measured will be in error.

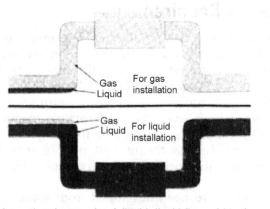

Figure 7-2 To keep the pipe running full with liquid flow, place the meter lower than the piping. For gas, keep the meter higher than the piping.

Measurement Units

In the measurement of flow, output is desired in some unit of volume or of mass. Volume flow rate or totalized volume measured at line conditions does not represent a defined quantity without a definition of the base conditions; they cannot be combined with other volumes whose fluid and line conditions are different. However, volume flow rate or totalized volume corrected to base conditions is a definitive quantity which can be combined with other values also at base conditions. Values of volume and mass at base conditions are related to each other through density at line conditions (mass value divided by base density). The most common method used to measure flow rate or totalized flow is a volumetric meter to measure at line conditions and a densitometer or chromatograph to measure fluid density also at line conditions. Mass flow rate or totalized mass may be measured directly with a direct mass meter.

If fluid density varies by more than about ±0.15% during a totalization period, then the conversion must be performed on a flow rate-weighted basis, or the conversion should be calculated more frequently to obtain the desired uncertainty. Depending on the billing requirements, other factors such as the heating value of individual components (which may have individual price values) may have to be derived from these base flow measurements, or determined with a Btu chromatograph.

Proper base conditions are normally determined by plant/sales contract, government requirements or operating agreement.

Installation Requirements

Each meter has certain requirements necessary for achieving its measurement potential. The requirements vary between meters and types of meters; refer to the specific meter section, standards, and manufacturers for details. They are determined by standards organizations, users, manufacturers, and research laboratories. In each case, the performance characteristics of the meter depend on these requirements being met. Any lessening of the specifications means that performance is compromised, and specific calibrations in place should be run (Figure 7-3).

Most meters have been developed from some basic principles embodied in a prototype. These prototypes may go through several development iterations, but if they prove out in flow

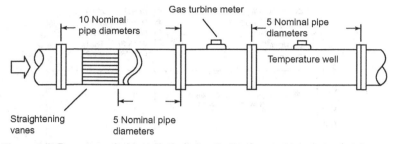
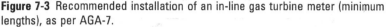

Figure 7-3 Recommended installation of an in-line gas turbine meter (minimum lengths), as per AGA-7.

tests, they become a marketable product. Except in special cases, most meters are designed for general usage and are tested accordingly. During these tests, effects on meter performance from flow into and out of the meter are determined and installation requirements are set. With sufficient experience in a given industry, plus additional tests, a standard may be prepared to guide users on the meter's installation requirements.

As time passes and additional applications are checked, these requirements may be adjusted to reflect new knowledge. Whatever the source, data should be checked to assure a user that the application has been evaluated and the design data are valid for the specific job requirements.

Flow Pattern

The heart of all this is the flow pattern entering the meter, which in general will be correct if the Reynolds number limit and the inlet piping agree with the original evaluation. The installation will then be proper. If a flow measurement practitioner deviates from the tested design, most meter manufacturers make no claim about their meter's performance and simply state that minimum piping requirements must be met.

This means two things are important: 1) familiarity with installation requirements is a must, and 2) the minimum requirements must be met to make sure no errors are introduced due to installation. Some meters have a defined flow pattern requirement (i.e., the amount of distortion of the flow pattern and/or swirl allowed), but this is rarely checked when a meter arrives in the field. Recent work has indicated that additional flow pattern preparation is necessary to reduce the inaccuracies in meters with heightened sensitivity to flow profile

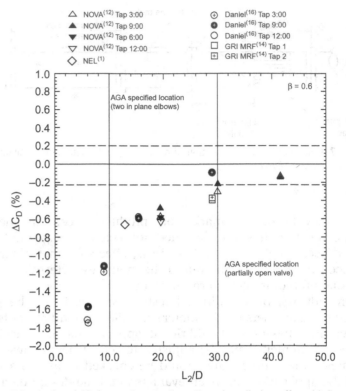

Figure 7-3a Isolated 90 degree long radius elbow installation effects' influence on a bare meter tube with 0.6 β ratio vs. upstream meter tube length.

variation. This has required changes in installation requirements for flange-tapped orifice meters as seen in ANSI/API 14.3/AGA Report. No.3, Part 2, to maintain proper flow patterns.

Much of the significant research into installation requirements has been conducted by the AGA-API and other international committees responsible for writing standards on orifices. The influence of installation effects can be seen in Figures 7-3a through 7-3d. These figures show the installation effects' influence on the orifice plate coefficient of discharge, Cd, from a relatively ordinary isolated long radius 90 degree elbow disturbance. Figures 7-3a and 7-3b show the influence on 17D upstream meter tubes without any flow conditioning (bare tubes) and two different diameter ratios (β), 0.6 and 0.4. Figures 7-3c and 7-3d show the influence on a 17D upstream meter tube with 19 tube uniform straightening vanes and two different diameter ratios (β), 0.6 and 0.4. The bare meter tube and straightening vane meter tube data show that upstream

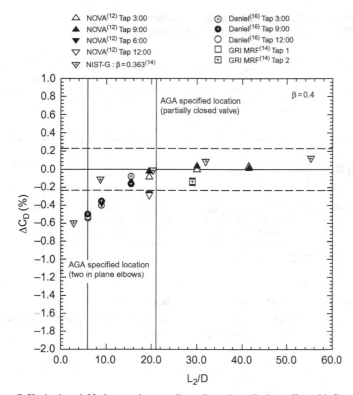

Figure 7-3b Isolated 90 degree long radius elbow installation effects' influence on a bare meter tube with 0.4 β ratio vs. upstream meter tube length.

meter tube lengths recommended in the 1992 standards were inadequate to prevent errors in the measurement, and these were revised in 2000. In addition the research data indicated that the installation effects' influence could be minimized in existing meters by reducing the diameter ratios (β) to 0.4.

A limited amount of work has also been conducted with linear meters, mostly gas and liquid turbine meters and ultrasonic meters.

From a design standpoint, the best interpretation of these revised requirements is to use the worst case (most disturbed profile) and design to it using the highest Reynolds numbers (the largest beta ratio) so meter tubes will be of a universal length that is independent of actual piping upstream or downstream. The result will be meter tubes which are longer than required by the standard for a particular application, but the standards provide *minimum* requirements, and longer meter tubes are better if space permits. The same philosophy applies

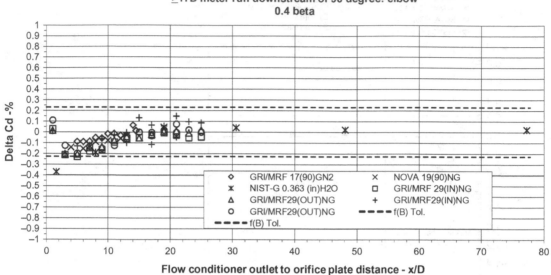

Figure 7-3c Isolated 90 degree long radius elbow installation effects' influence on a meter tube with straightening vane using 0.6 β ratio vs. upstream distance between vane outlet and orifice plate.

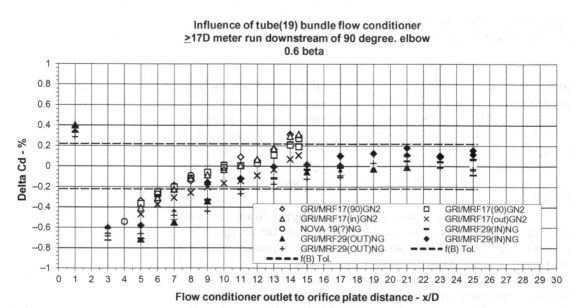

Figure 7-3d Isolated 90 degree long radius elbow installation effects' influence on a meter tube with straightening vane using 0.4 β ratio vs. upstream distance between vane outlet and orifice plate.

to meter tubes for any type of meter where upstream/downstream piping is a consideration.

Flow pattern distortion comes from upstream pipe fittings, foreign material in the meter tube, or improperly made or aligned meter tubes. Any of these will cause an asymmetrical flow pattern with the possibility of swirl. Asymmetrical profiles can cause errors in the 0–3% range with differential meters and ultrasonic meters, but errors are generally less for intrusive linear meters such as turbine and positive displacement (PD). Errors from swirl may be in the 1–6% range for differential meters and generally more (1–10%), depending on swirl direction and magnitude, for other types of meters

Obviously, swirl is the major concern. If there is any question of swirl being present, straightening vanes should be used. Experience has indicated that swirl can propagate even in extra-long (250 nominal pipe diameters) upstream piping. Tank/vessel outlets and parallel meter tube headers are known to generate 10 to 50 degrees swirl angle magnitude. The better design is to minimize changes in direction and planes of flow upstream of the meter. If swirl is not generated, it obviously will not be propagated (Figure 7-4).

In addition to generating swirl, upstream meter piping can create or minimize other problems. High velocities across blocked in-line tees can create vortices in both the tee and the line. This can set up fluctuations or pulsations that make flow metering difficult. On gas lines that may drop out liquids, pockets of liquids can also create variable flow rates as the fluids wash back and forth. Drips with drains should be put in low spots to drain this liquid away (Figure 7-5).

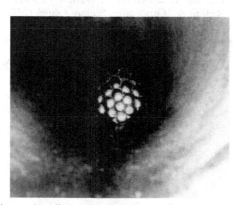

Figure 7-4 Swirl must be eliminated for accurate flow measurement. Note swirl pattern on pipe wall.

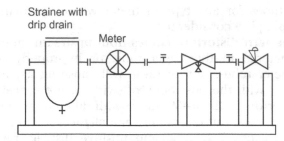

Figure 7-5 Drips help keep gas lines free of liquids, and air eliminators keep liquid lines free of gas.

If pipelines are very dirty, it may be necessary to install scrubber/filters to filter fluids and solids out of gases, and solids and gases out of liquids. On liquid lines, where gas or vapor may be present, air/vapor eliminators should be installed or flow fluctuations may result. Sizing of upstream headers should be controlled so that velocities are slowed in the lines leading to meter runs.

A useful rule of thumb is that the area of the header should be one-and-a-half to two times the area of all of the meter runs off it. An alternate rule is to have velocity in the headers half of that in the meter tubes. This will provide good flow distribution to the meters and minimize flow profile problems caused by the headers.

References

American Gas Association, 1515 Wilson Boulevard, Arlington, VA 22209.

API, Manual of Petroleum Measurement Standards. Chapter 14, "Natural Gas Measurement," Section 3, "Concentric, Square-Edged-Orifice Meters," Parts 1−4.

International Organization for Standardization (ISO), Case Postale 36, CH 1211 Geneva, Switzerland. Or in the United States: ANSI (for ISO Standards), Customer Service, 11 West 42nd St., New York, NY 10036, phone: (212) 642−4900.

8

OPERATIONS

Operational Considerations

After examining flow and fluid conditions, choosing the correct meter type for a specific application is the next step in achieving minimum measurement uncertainty. However, the meter's limitations must be recognized along with its positive features in order to make the best selection. Most meters operate with a specified uncertainty within stated flow capacity limits. For custody transfer applications, a meter should not be operated for extended periods of time at or below its stated minimum flow or above its stated maximum capacity.

The primary consideration for custody transfer measurement is to minimize flow variations by maintaining better control of the flow rate. At times, this may not be possible, and the need for a meter with a wide ranging flow capacity will be a further consideration in its selection. If a single meter with the required flow capacity to cover the intended operating range with minimum uncertainty does not exist, the use of multiple meters with some type of meter switching control is required.

For example, consider the fuel supplying a process with three heat exchangers. The range of fuel flows required may be from the pilot load to all three exchangers in full load service. This could require a measurement flow range of over 100 to 1. The metering used could be a combination of a positive displacement meter for the low flows and several turbine or orifice meters for the high flows. Other meter types might also be used. At one time, implementing this type of complex metering system with its meter switching controls could have been a problem, but computers and electronic controllers have simplified the situation, and make such multiple metering solutions more practical. Such a metering system can now be easily managed, and flows can be accurately measured, with the total flow for the whole station being reported as a single measurement.

In addition to the problems of operating meters at the extremes of their flow capacities, the secondary equipment that

Fluid Flow Measurement. ISBN: 978-0-12-409524-3

measures pressure, temperature, differential pressure, density or relative density (specific gravity), and flow composition can also have limitations. Typical uncertainty specifications for these devices are stated as a percent of full scale. Selecting an instrument with the wrong range for the parameter to be measured may introduce greater uncertainty.

If the flowing pressure to be measured is 75 pounds per square inch gauge (psig), and it is measured by an instrument with a 1,000 psig range that has a ±0.5% uncertainty at full scale, then the pressure measurement error could be as high as 5 pounds out of 75 or ±6.7% for linear meters and ±3.3% for differential meters. (The reason for the difference in the two values is that the pressure directly influences the linear meter calculation, but its influence on the differential meter's calculation is decreased due to the square root extraction.)

The best operating range for a metering system is between 25 and 95% of the maximum capacity of the meters. If operational changes do not exceed the range of the meter, it should be selected to operate near its maximum capacity (Figure 8-1).

Operational Influences on Gas Measurement

Table 8-1 is based on the measurement of natural gas with an orifice meter. The exact magnitude of errors and dollars is not as important as realizing that significant financial risk is at

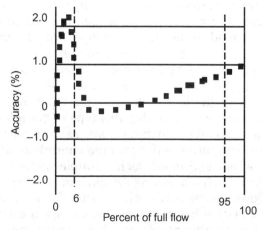

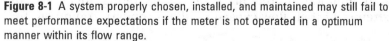

Figure 8-1 A system properly chosen, installed, and maintained may still fail to meet performance expectations if the meter is not operated in a optimum manner within its flow range.

Table 8-1 Custody Transfer Measurement

Flowing Temperature Error Temperature, °F	Flow, Mcfd	Loss/Day	Loss/Year
2.0	3,837		
1.8	3,641		
−0.2 iwc Error	196	$784	$286,160
25.0	13,553		
24.8	13,499		
−0.2 iwc Error	54	$216	$78,840
90.0	25,678		
89.8	25,650		
−0.2 iwc Error	28	$112	$40,880

Reflects an error in differential pressure measurement of −0.2 in.

Static Pressure Error Pressure, psia (Differential of 2.0″)	Flow, Mcfd	Loss/Day	Loss/Year
600.00	3,837		
598.00	3,831		
−2 psi Error	6	$24	$8,760
(Differential of 25.0″)			
600.00	13,553		
598.00	13,528		
−2 psi Error	25	$100	$36,500
(Differential of 90.0″)			
600.00	25,678		$112	$40,880
598.00	25,632		
−2 psi Error	46	$184	$67,160

Reflects an error in pressure measurement of −2 psi for differential meter flow rates.

Flowing Temperature Error Temperature, °F (Differential of 2.0″)	Flow, Mcfd	Loss/Day	Loss/Year
62.0	3,828		
60.0	3,837		
+2°F Error	9	$36	$13,140
(Differential of 25.0″)			
62.0	13,519		
60.0	13,553		
+2°F Error	34	$136	$49,640

(Continued)

Table 8-1 (Continued)

Flowing Temperature Error Temperature, °F (Differential of 90.0")	Flow, Mcfd	Loss/Day	Loss/Year
62.0	25,615		
60.0	25,678		
+2°F Error	63	$252	$91,980

Reflects an error in temperature of +2°F for differential meter flow rates.

stake and proper operation is vital. A similar study should be performed for any custody transfer application.

Natural gas prices have risen significantly since it was first commercialized, and prices as high as $15 per Mcf have been experienced. Table 8-1 shows the influence of small errors on calculated volumes. Although the examples are based on errors due to instrumentation reading low, similar calculations can be made for instrumentation reading high. The calculations are based on a single 8 inch meter tube using a 4.000 inch bore orifice plate and a gas with a relative density (specific gravity) of 0.580 and 0% carbon dioxide and nitrogen. Volumes are calculated using differential pressures of 2.0, 25, and 90 inches water column (iwc), static pressure of 600 psia, and flowing temperature of 60°F to show the monetary impact of small calibration and/or operating errors. A price of $4/Mcf was used to show revenue errors.

As noted, each meter type has its optimum area of operation for achieving minimum uncertainty, and meters must be matched to the rangeability of the expected flows to be measured. If the flows are steady day in and day out, a meter with a limited rangeability may be all that is needed to handle the application. However, most flow rates in the oil and gas industry change continually. Selecting the metering to accommodate varying flow rates with minimum uncertainty then becomes a major consideration in the design and operation of a meter station. Most meters encounter greater uncertainty in the lower 0 to 10% of their flow capacity. Therefore, if the flow range to be measured includes this area of operation, appropriate designs, using multiple meters, expanded readout systems, and/or characterization of the meters must be employed.

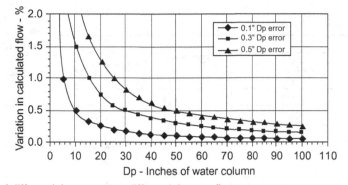

Figure 8-2 Influence of differential pressure on differential meter flow.

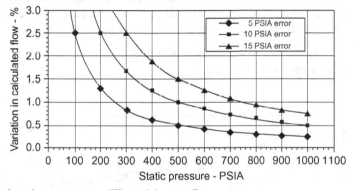

Figure 8-3 Influence of static pressure on differential meter flow.

Another way to look at the influence of the flowing conditions on uncertainty is to plot "variation" (error) versus the parameter as in the four charts in Figures 8-2—8-5.

A common misconception is that all meter types maintain the same uncertainty over their entire flow capacity. This tends to be a good assumption for linear meters, but is not true of differential meters. Linear meters normally have an uncertainty that is stated as a percentage of flow or reading, whereas differential meters normally have an uncertainty that is stated as a percentage of full scale or maximum capacity.

For example, a turbine meter has a flow uncertainty expressed as a percentage of flow rate. The uncertainty statements for temperature, pressure (on both linear and differential meters) and differential pressure for differential meters are stated as a percentage of full scale or maximum calibrated span. To summarize this subject, uncertainties in metering are stated in one of two ways: percentage of actual flow rates or reading; or percentage of maximum capacity or full scale.

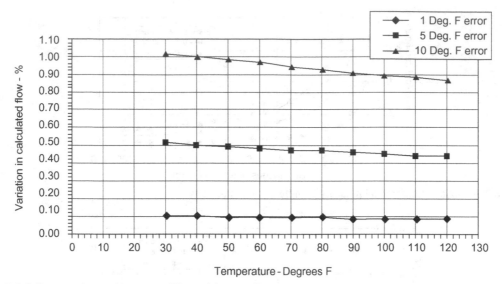

Figure 8-4 Influence of temperature on differential meter flow.

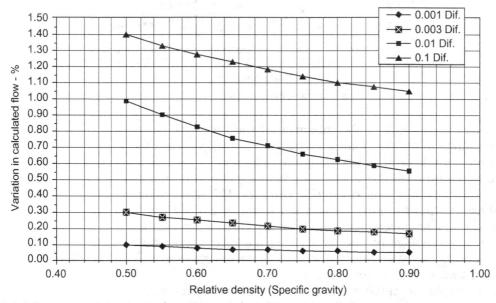

Figure 8-5 Influence of relative density (specific gravity) on differential meter flow.

To make a proper comparison of the "uncertainty" numbers, the statement of each meter's uncertainty with all operating limitations must be known. These numbers guide an operator in selecting the meter's optimum capacity range in order to obtain minimum uncertainty.

For a meter with a percentage of flow rate uncertainty statement (such as turbine or ultrasonic), the uncertainty is the same over its entire stated capacity range. The uncertainty of "percentage of maximum capacity" meters can be directly compared to the uncertainty of "percentage of flow rate" meters only at their maximum capacities. Below maximum capacity or full scale, the percentage uncertainty increases as the flow rate decreases for "percentage of maximum capacity" meters, while for "percentage of flow rate" meters uncertainty remains unchanged. Therefore, the two meters' uncertainties are not directly comparable at lower flow rates.

The significance of flow rate differences on uncertainty is now apparent. When metering systems consist of multiple meters for flow measurement, the uncertainty of all transmitters must be considered to estimate overall system uncertainty. It is very important that each transmitter's uncertainty be stated in the same terms to make a valid statement of the calculated overall system uncertainty. Likewise, point calibrating a transmitter in a narrow range of operation (such as temperature and pressure) may produce less uncertainty over a limited range than the manufacturer's stated overall uncertainty.

Most gas metering systems with properly selected, installed, operated, and maintained meters and transmitters should achieve measurement uncertainties under actual operating conditions in the range of ±1%. However, improperly applied meters can misinform a user who does not have a complete knowledge of their operational limitations.

Uncertainty

An estimation of the flow metering performance under given operating conditions can be made with an uncertainty calculation. Many uncertainty calculation procedures are available in the industry standards and flow measurement literature, such as ANSI/ASME MFC-2M, "Measurement Uncertainty for Fluid Flow in Closed Conduits." The value of an uncertainty calculation is not so much in the absolute value obtained, but rather in the relative comparison of the overall system to the uncertainty of other meters or metering systems, and in exploring the sensitivity of individual components—differential pressure, static pressure, flowing temperature, etc.—in the measurement system.

Such calculations must utilize the particular operating conditions for a specific application in order to be most useful in obtaining minimum measurement uncertainty.

Equation 8.1 shows the calculation of uncertainty for an orifice meter. It is, therefore, representative of differential meters in general and also shows the manner in which the calculation can be made for any meter. The equation for the calculation is divided into two types of error—bias (B) and precision (S)—then these are combined as the square root of the sum of the squares.

$$U = \sqrt{(B_1{}^2 + B_2{}^2 + \ldots) + (S_1{}^2 + S_2{}^2 + \ldots)} \qquad (8.1)$$

This equation will calculate the uncertainty of a flow measurement, assuming the variables are measured over a long time period. The uncertainty represents deviation from the true value for 95 percent of the time. The value of the calculation will depend on the accuracy of the values used for the bias and precision errors. Individual component uncertainties are weighted, based on the manner in which they influence the flow calculation. In the simplified orifice equation:

$$Q = Kd^2 \times (dP \times P)^{1/2} \qquad (8.2)$$

where:

Q = rate of flow in appropriate units;
K = a coefficient based on the mechanical installation and other flow variables;
d = orifice bore diameter in appropriate units;
dP = differential pressure in appropriate units;
P = absolute static pressure in appropriate units.

The values are either direct multipliers, squared, inverse, or square root values. The weighting values have a sensitivity factor of either 1 or -1 for direct multiplied variables, a sensitivity factor of 2 for squared value, and a sensitivity factor of 1/2 for square root value multiplied by component uncertainties. Thus, in the simplified orifice meter uncertainty Equation 8.2, the uncertainty in K is multiplied by 1, the uncertainty in d is multiplied by 2, and the uncertainty in dp and P are multiplied by 1/2. Thus, the uncertainty in orifice meter flow rate would be:

$$U = \sqrt{B_k^2 + (B_d)^2 + (\tfrac{1}{2}B_{dP})^2 + (\tfrac{1}{2}B_P)^2 + S_k^2 + (2S_d)^2 + (\tfrac{1}{2}S_{dP})^2 + \tfrac{1}{2}(S_p)} \quad (8.3)$$

The values used for the precision uncertainties in the equation may be obtained from the manufacturers' specifications for the respective pieces of equipment, provided that the values are adjusted to reflect operating conditions. The bias uncertainties must be determined by testing.

Examples of Gas Differential Meter System Uncertainties

Since there are many combinations of equipment, operating conditions and calculation methods in existence for orifice metering, it is impossible to establish a single base line uncertainty relationship. The most practical approach is to provide uncertainty ranges for the most typical differential metering combinations. The following orifice metering system combinations have been selected as examples of system uncertainty estimations (Table 8-2):

- Chart metering system (upper range) operating under AGA Report No. 3 (1985) with differential pressure averaging 10 inches of water column (iwc), single static pressure and temperature, and standard calibration equipment accuracy.
- Chart metering system (lower range) operating under AGA Report No. 3 (1990 to 1992) with differential pressure averaging 50 iwc and premium calibration equipment accuracy.
- Electronic metering system (upper range) operating under AGA Report No. 3 (1985) with differential pressure averaging 10 iwc, single static pressure and temperature, and standard calibration equipment accuracy.
- Electronic metering system (lower range) operating under AGA Report No. 3 (1990 to 1992) with differential pressure averaging 50 iwc and premium calibration equipment accuracy.
- For the purpose of establishing the orifice meter measurement uncertainty ranges, the following conditions are assumed:
  - The variables P_f, T_f, and G_r are functioning at 70% of full scale.
  - Ambient temperature influences are maintained at $\pm 15^\circ$F through monthly calibrations.
  - The differential pressure variable, dp, is maintained between 10 and 95% of full scale.
  - All meter runs are of the same size with the same bore orifice plates.
  - There is an equal distribution of flow among meter tubes.

Figures 8-6 and 8-7 provide the results of the chart and electronic system multiple meter tube orifice meter uncertainty calculations.

Example of Gas Linear Meter System Uncertainties

Since there are numerous combinations of equipment, operating conditions, and calculation methods in existence for linear metering, it is impossible to establish a single uncertainty

Table 8-2 Orifice Meter System Uncertainty Examples

Element	Chart System Element % Accuracy Upper Range	Chart System Element % Accuracy Lower Range	Electronic System Element % Accuracy Upper Range	Electronic System Element % Accuracy Lower Range
Cd	0.600	0.440	0.600	0.440
Y	0.030	0.010	0.030	0.010
d	0.050	0.020	0.050	0.020
D	0.250	0.010	0.250	0.010
dp	0.500	0.500	0.150	0.150
Pf	1.000	1.000	0.250	0.200
Tf	1.000	1.000	0.250	0.100
Fpv	0.100	0.100	0.100	0.100
Gr	0.500	0.100	0.500	0.100
d_{pc}	0.100	0.050	0.100	0.050
P_{fc}	0.100	0.050	0.100	0.050
T_{fc}	0.067	0.067	0.067	0.067
G_{rc}	0.167	0.167	0.167	0.167
	Single Meter Run Calculated % Uncertainty **2.85**	Single Meter Run Calculated % Uncertainty **1.20**	Single Meter Run Calculated % Uncertainty **1.19**	Single Meter Run Calculated % Uncertainty **0.60**

Where:
Orifice meter coefficient of discharge, Cd
Expansion factor, Y
Orifice bore diameter, d
Meter tube inside diameter, D
Differential pressure, dp
Static pressure, Pf
Flowing temperature, Tf
Gas compressibility factor, Zf & Zb (Fpv)
Gas relative density, Gr
Differential pressure calibrator, d_{pc}
Static pressure calibrator, P_{fc}
Flowing temperature calibrator, T_{fc}
Gas relative density calibrator, G_{rc}

relationship. The most practical approach is to provide uncertainty ranges for the most typical linear metering combinations.

The following have been selected as the most typical linear metering combinations (Table 8-3):

- Chart metering system (upper range) with linear meter operating at less than 1% of capacity, single static pressure and temperature, and standard calibration equipment accuracy.

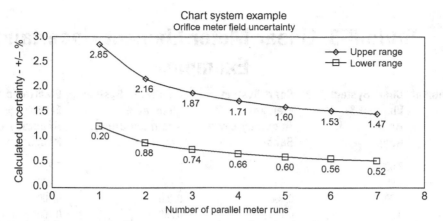

Figure 8-6 Example of the range of orifice meter uncertainty using circular charts.

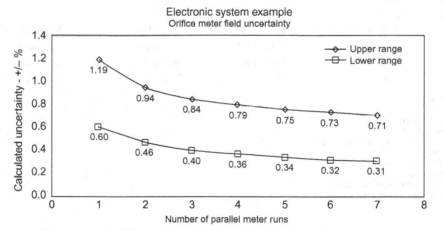

Figure 8-7 Example of the range of orifice meter uncertainty using the electronic system.

- Chart metering system (lower range) with linear meter operating from 10 to 100% of capacity and premium calibration equipment accuracy.
- Electronic metering system (upper range) with linear meter operating at less than 1% of capacity, single static pressure and temperature, and standard calibration equipment accuracy.
- Electronic metering system (lower range) with linear meter operating from 10 to 100% of capacity and premium calibration equipment accuracy.

Table 8-3 Linear Meter Element Uncertainty Examples

Element	Chart System Element % Accuracy Upper Range	Chart System Element % Accuracy Lower Range	Electronic System Element % Accuracy Upper Range	Electronic System Element % Accuracy Lower Range
PM_L	2.000	0.500	2.000	0.250
Pf	1.000	0.500	0.250	0.200
Tf	1.000	0.500	0.250	0.100
Zf	0.250	0.100	0.250	0.100
Gr	0.500	0.100	0.500	0.100
PM_{fc}	0.500	0.300	0.500	0.300
P_{fc}	0.100	0.050	0.100	0.050
T_{fc}	0.067	0.067	0.067	0.067
G_{rc}	0.167	0.167	0.167	0.167
	Single Meter Run Calculated % Uncertainty	Single Meter Run Calculated % Uncertainty	Single Meter Run Calculated % Uncertainty	Single Meter Run Calculated % Uncertainty
	3.24	**1.36**	**2.58**	**0.73**

Where:
Positive meter linearity, PM_L
Static pressure, Pf
Flowing temperature, Tf
Gas compressibility factor, Zf & Zb
Gas relative density, Gr
Positive meter flow calibrator, PM_{fc}
Static pressure calibrator, P_{fc}
Flowing temperature calibrator, T_{fc}
Gas relative density calibrator, G_{rc}

For the purpose of establishing the linear meter measurement uncertainty ranges, the following conditions are assumed:

- The variables Pf, Tf, and Gr are functioning at 70% of full scale. Ambient temperature influences are maintained at $\pm15°F$ through monthly calibrations.
- All meter runs are of the same size with the same size meters.
- There is an equal distribution of flow among meter tubes.

Figures 8-8 and 8-9 provide the results of the chart and electronic system multiple meter tube linear meter uncertainty calculations. These are "lost and unaccounted for" control charts.

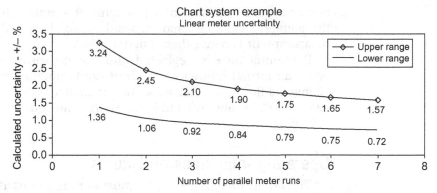

Figure 8-8 Examples of the range of linear meter uncertainty using circular charts.

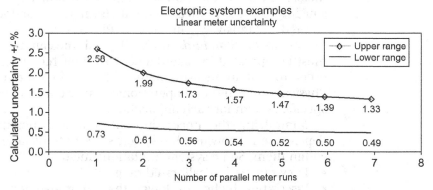

Figure 8-9 Examples of the range of linear meter uncertainty using the electronic system.

The foregoing is a somewhat detailed description of the influence of only one of the factors in determining uncertainty from the equation. Many other values should be similarly examined.

Calculation of the uncertainty associated with the variables in the flow equation is not the only concern for a complete uncertainty determination. Allowance must be made for human interpretation influence, chart integration or computer influence, installation influence, and fluid characteristic influence. Most of these influences are minimized, provided industry standard requirements are met and properly trained personnel are responsible for the operation and maintenance of the station. Since these effects cannot be quantified, they are minimized by recognizing their potential existence and properly controlling the meter station design, operation, and maintenance. Without proper attention to the total problems, a simple calculation of the equation variables may mislead a user into believing that measurement is better than it actually is.

If maintenance is neglected and the measurement experiences abnormal influences, such as contaminant deposits that change its flow characteristics, then the calculation is meaningless until those abnormal influences are removed.

Operating Influences on Liquids

The liquid metering systems most commonly used in the oil industry are turbine metering, positive displacement (PD) metering, and tank gauging. Additionally, the use of ultrasonic and Coriolis meters is growing rapidly. Turbine meters and PD meters measure the flowing stream dynamically, while tank gauging is a static measurement. Both types of metering are covered extensively in the API *Manual of Petroleum Measurement Standards* (MPMS), and measurement systems must be operated in accordance with their recommendation in order to obtain measurement with a minimum uncertainty. Physical properties and operational constraints will determine a specific system for each application.

Minimizing the uncertainty of a liquid metering system depends on knowing the system's limitations and operating within them. Some system limitations include:

- Flow rate within calibrated range;
- Viscosities higher or lower than that experienced during calibrations;
- Operating at extreme flow rate; and
- Operating at extreme temperatures.

For turbine and PD meters, a general review should include:

- Is the station actually operating within the design flow range?
- Is the system designed to take into account all of the physical properties, such as temperature, pressure, density or relative density, and to be compatible with any corrosive characteristic of the fluid to be measured?
- Is the meter protected from excessive operating conditions, such as liquid surges, entrained gases (flashing), pulsations, and excessive pressure, and are the protective devices such as pressure relief, surge tanks, and vapor removal equipment installed and operating properly?
- Is the flowing environment clean (especially important for turbine meters)?

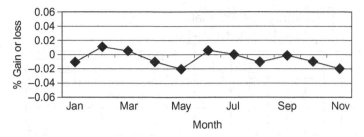

Figure 8-10 Control chart.

- If the liquid has a high vapor pressure, is back pressure monitored and controlled?

The influence of these conditions can be identified and tracked by the use of an operating system control chart such as shown in Figure 8-10.

The value of good measurement should be obvious to the readers of this book. To obtain this quality of measurement there are two main areas affecting meter operation: system parameters and meter parameters.

System parameters should include fluid viscosity. Both turbine meters' and positive displacement meters' performance can be affected by liquid viscosity. A PD meter is more linear than a turbine meter at higher viscosities. As viscosities increase, a turbine meter's meter factor can become significantly non-linear. On the other hand, PD meters tend to be less sensitive to viscosity variations and more concerned with slippage and pressure drop. At lower flow rates, a PD meter factor curve can be affected as viscosity drops due to increased slippage. As viscosities go up, pressure drop increases and meter wear is increased. If the liquid viscosity changes widely, meter performance will be affected by the deviations as shown in the previous charts. Examples of the influence of varying viscosity on the performance of PD meters and two types, conventional and helical, of turbines meters are shown in Figures 8-10a, 8-10c, and 8-10e. In addition, Figures 8-10b, 8-10d, and 8-10f show how the performance of these meters can be improved by characterization using the information from meter factor control charts.

When searching for sources of increased meter uncertainty (particularly in liquid applications), it is necessary to investigate viscosity changes and correlate them with meter performance.

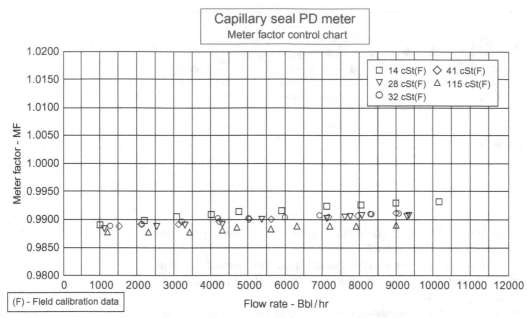

Figure 8-10a Meter factor control chart for a capillary seal PD meter experiencing varying viscosities.

Other Fluid Flow Considerations

Liquid flow rate can affect meter uncertainty. When examining a metering system, it is necessary not only to examine the total volumes measured, but also to know how any variations in meter flow rate influence the uncertainty in the total volume delivered.

The influence of temperature fluctuations must also be considered. In addition to affecting viscosity, liquid temperature can also change the mechanical clearances in PD meters and the measuring chamber volumes; changes in mechanical clearances may necessitate the use of extra clearance rotors. Additional mechanical clearance allows for different thermal expansions of the rotor and housing, and prevents meter lockup.

When proving meters, temperature equilibrium should be maintained throughout the meter and prover system. Where extremes of ambient or flowing fluid temperatures are experienced, it may be necessary to insulate the system to obtain stabilization. Temperature stabilization may require some run time before proving is attempted, since unstable temperatures will usually result in erratic provings.

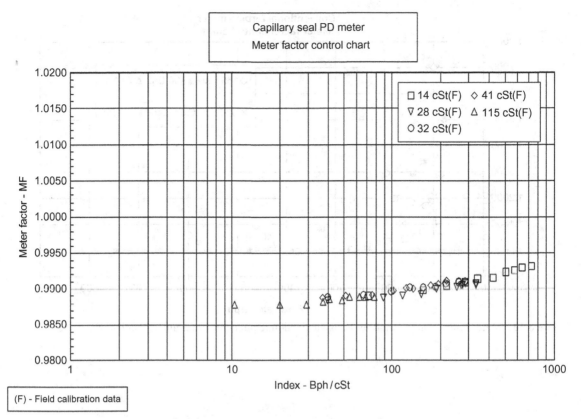

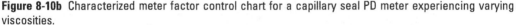

Figure 8-10b Characterized meter factor control chart for a capillary seal PD meter experiencing varying viscosities.

System flow rate must be known so that the meter range can be checked. Rating a meter so it operates in its mid-range (10 to 95% of the meter's stated range) will usually result in minimum uncertainty.

The nature of the liquid to be metered is critical to the selection of meter type. Metal in contact with flowing fluids must be compatible with the characteristics of the liquid to be measured. The liquid should be free of entrained air and abrasive solids. To eliminate air, an air eliminator can be installed upstream of the meter. It should be tested to insure proper operation. Likewise, a separator with properly sized mesh should be installed to remove the solids, and this should be checked periodically. Air that has not been removed will be measured as liquid. A slug of air followed by liquid into a turbine meter can destroy the turbine rotor and/or its bearings.

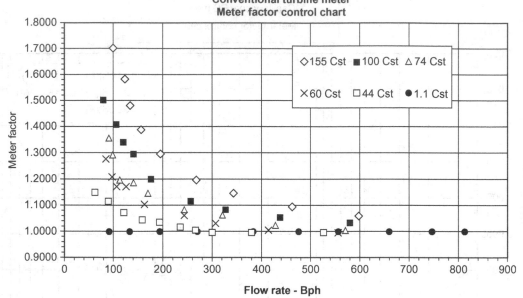

Figure 8-10c Meter factor control chart for a conventional turbine meter experiencing varying viscosities.

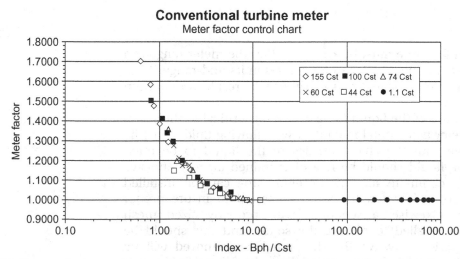

Figure 8-10d Characterized meter factor control chart for a conventional turbine meter experiencing varying viscosities.

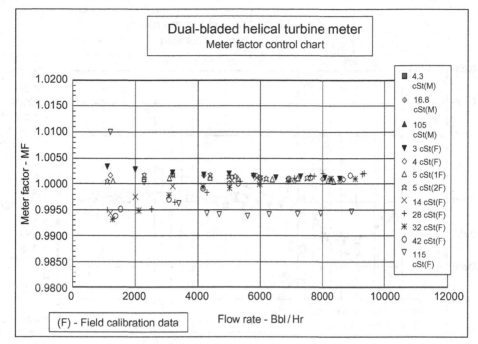

Figure 8-10e Meter factor control chart for a helical turbine meter experiencing varying viscosities.

Meter Parameters

The most important parameter in a meter's operation is its proving results. For both PD and turbine meters, changes in the meter factor curve outside of acceptable tolerances are an indication that an effective change is taking place, which may signal the need to disassemble the meter to repair any damage. The meter provings should be summarized in a meter factor control chart that can be monitored to identify changes in meter performance.

In general, a turbine meter's factor begins to become nonlinear at flow rates equal to, or below, 5% of the meter's maximum capacity. This is a point where bearing friction can become significant. With higher viscosities, the point of deviation may be as high as 35% of maximum capacity. Turbine meters are normally monitored electronically, so any losses due to readout should be insignificant. Once a base meter factor curve has been established for a meter, changes outside established bounds will indicate cleaning or replacement is needed.

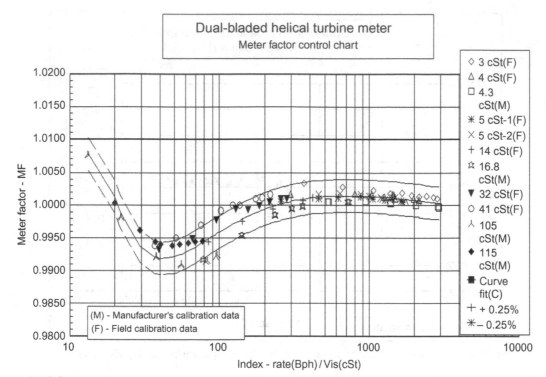

Figure 8-10f Characterized meter factor control chart for a helical turbine meter experiencing varying viscosities.

Rangeability

PD meters have a usable flow range of between 50:1 and 100:1. The turbine meter has about a 10:1 range with low viscosity liquids or with viscosity characterization (indexing). A turbine meter's usable range decreases at higher viscosities without indexing and with decreasing relative density (relative density of 0.55 or less).

Repeatability

The repeatability of a properly applied PD meter should be in the range of ±0.05% or better with the same flowing parameters. Turbine repeatability should be ±0.02% or better when the meter is operating properly. Unless a turbine meter is mechanically impaired, it will maintain its repeatability even when the meter becomes non-linear.

Pressure Drop

As previously indicated, pressure drop is affected by many factors including:

Flow rate;

Viscosity;

Density or relative density;

Temperature;

Flowing conditions.

Any significant change in pressure drop will signal that inspection and/or repairs should be scheduled as soon as possible. Typically, the pressure drop on a turbine meter is about 4 psi at maximum flow (6 psi at 130% flow) on water and is a function of each meter's internal design. At higher viscosities, the pressure drop will be larger. Pressure drop does not pose the same potential problem for a turbine meter that it does for a PD meter.

Output Signal

The output signal may be obtained through mechanical gearing systems or via an electronic pickup that produces a pulse signal. Turbine meters normally use an electronic readout.

Summary

There are a number of system and meter parameters that affect meter performance. By being aware of these concerns, a user can obtain the optimum performance from a meter that is maintained and operated properly. Installing a meter with excellent performance potential without providing for its proper operation and maintenance represents waste and incompetence.

Tank Gauging

In tank gauging, the most important parameters to measure correctly are:

Level readings;

Specific gravity (relative density);

Temperature;

Free water and sediment;

Tank dimensions.

Proper level measurement, correct tank dimensions, and proper temperature are all critical. Since temperature stratification may take place in a tank, getting a representative average temperature may be difficult. Several procedures are specified and must be used in accordance with the appropriate types of vessel, such as storage tank, tank cars, and trucks. Another concern is wall retention, i.e., the amount of the fluid that may adhere to the wall within a tank.

Tank strappings on all vessels should be up to date. A tank should have a solid foundation under its floor, and the walls should not be flexible. The walls should be clean and not encrusted.

The general rule of thumb is that the best uncertainty (0.15%) obtained when using tank gauging occurs when the entire volume of a completely full tank is withdrawn. When a very small amount (in the order of 4%) is withdrawn from a 100% full tank, the uncertainty suffers and can be in the order of 3%. Again when the entire volume of a 15% full tank is withdrawn the uncertainty will suffer and can be as great as 0.66%.

It Pays to Protect Investment

Operating flow measurement problems are solved by dedicating people, time, and money to ensuring that the meters and equipment are operating properly. Delivery variances are thus minimized and terminals and pipelines inventories are controlled. To aid in finding problems in the case of an imbalance, documentation should be maintained that shows that the calculations, procedures, and equipment operation have been performed in accordance with the industry standards and contract requirements.

Some flows are difficult to measure accurately with any meter. Once these are identified, the operator must accept greater uncertainty in the measurement, or should improve the flowing conditions to allow more accurate measurement. It is imperative that any meter's limitations be recognized and fitted to the operating requirements of the flows to be measured. The performance potential of a meter is of no value if it is not installed, operated, and maintained in such a manner that its potential can be realized.

9

MAINTENANCE OF METER EQUIPMENT

Both the shipper and the receiver must have confidence that a custody transfer meter is measuring the proper delivery volumes and meeting contract requirements. Equipment calibration may change over time, so *both parties* should take an active part in the periodic testing of the meter system. Without tests to reconfirm the original accuracies, a statement of accuracy is not complete and may be misleading.

Maintenance tests usually depend on contractual requirements for type and frequency—as often as weekly or even daily. They may require only a calibration of the readout equipment, a complete mechanical inspection of the entire system, or an actual throughput test against some agreed-upon correct volume.

Maintenance testing may consist of only secondary equipment calibration, or may involve a complete mechanical inspection of the entire system, or an actual throughput test against some agreed-upon standards—or any combination of these.

In any case, the equipment that is used to test the meter must be approved and agreed upon. Such test devices include certified thermometers for temperature, certified dead-weight testers or test gauges for pressure, certified differential testers for differential meters, certified chromatographs for component analysis, and certified provers for throughput tests. Many models of each are available and can be supplied with accuracy certification. Certification is important for both parties to minimize concern about the test equipment acceptability.

Operators who have had experience with similar metering systems will increase the confidence level in the calibration equipment and test procedures. The test equipment itself should be recertified on a timely basis by the agency or manufacturer that originally certified the equipment.

The first step in testing any meter is a visual inspection for any signs of improper operation, such as leakage and

Fluid Flow Measurement. ISBN: 978-0-12-409524-3

unstable flow. This includes a review of all of the attendant equipment and their indications or recordings. If the station appears to be operating properly, the individual elements of the station, such as the meter and the corrections for pressure, temperature, density, and composition, should be individually verified and/or calibrated with the assumption that if all parts are in calibration, the system will be in calibration to the limits calculated by the uncertainty equation. This procedure is commonly used for industrial flow metering.

A master meter used in **transfer proving** is calibrated and certified to some uncertainty limit by a testing facility of a government agency, a private laboratory, a manufacturer, or the user using agreed-upon flow standards. Periodically, the master meter has to be sent back to the laboratory for recertification. The frequency of this retesting depends on the fluids being tested and the treatment of the master meter between tests.

The best throughput test is one that can be run directly in series with a **prover**. The prover can come in many forms, but essentially it involves a basic volume that has been certified by the government or an industry group. Such provers for liquid may be calibrated Seraphin cans, certified field test measures, (for fluids with no vapor pressure at flowing temperature), pressurized volume tanks (for fluids with vapor pressure at flowing temperature), or pipe provers (formerly called mechanical displacement provers as described in API's *Manual of Petroleum Measurement Standards*). These pipe provers are permanently installed in large-dollar-volume meter stations, but portable units for testing smaller meter stations also exist.

Gas provers are usually master meters with computer controls so that testing requires little or no calculation, or critical flow nozzles (where accurate thermodynamic properties of the gas are available). Critical nozzles-venturi require permanent pressure drops of some 15 to 20% of the upstream static pressure and cannot be run at static pressure below approximately 30 pounds per square inch absolute.

Atmospheric critical flow proving, which requires gas to be exhausted to the atmosphere, used to be popular for linear meters. This is now seldom used because of the cost of lost gas and concerns about safety and the environment.

Great care must be exercised in using such equipment, as detailed by standards or manufacturers' instructions, to ensure accurate testing. Since these tests are subject to errors, only qualified technicians should conduct them. Run correctly, the tests ensure the best measurement and provide proof of uncertainty (Figure 9-1).

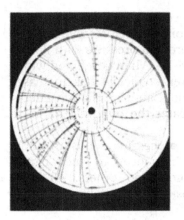

Figure 9-1 A major problem with tests using charts—such as this one comparing chart integration versus a flow computer—is that even though input data are correct, data extraction by integration may introduce *very* large errors.

It should be noted that testing requires participation by both the supplier and the customer. Diagnostics and evaluation with proper test equipment ensure that recorded volumes are correct. Any proving must be documented and signed by both parties so that contract provisions can be implemented on any corrections required.

Gas Measurement Maintenance

In a rush to "re-engineer" pipelines—improve profits by lowering cost—a favorite target has been the maintenance costs along gas pipelines. Because of the reduction in personnel and time allocated for testing and maintenance, quality maintenance time has been traded for meeting a schedule of routine work and meeting emergencies. This has resulted in less control over the quality of gas transported and the accuracy of the meters used to measure it.

Maintenance for all meters, particularly orifice and turbine for gas measurement, has suffered. In order to avoid having to change orifice plates to accommodate changing flow ranges, the largest orifice bore acceptable to both parties is often installed along with smart differential pressure transmitters.

Those responsible mistakenly believe such meters can operate without maintenance for greatly extended time periods at *very low differential pressures (less than 10% of upper range value or 10 inches water column whichever is greater)* because the smart transmitters have been calibrated at low ΔP with acceptable uncertainties. This focus on low differential calibration misses the more important consideration about noise and the true ΔP in a dynamic system operating at low ΔP.

Research has clearly shown that volume variations in excess of 5% can occur under such conditions. Therefore, volume and/or energy determination at low differential pressure with flange-tapped orifice meters is inherently risky. When combined with a large beta ratio, the metering system typically has increased sensitivity to pulsation, contamination, and installation effects. The poorer the gas quality, the greater the potential for measurement error.

Measurement upstream of gas processing tends to suffer greater uncertainty and is greatly influenced by such factors as particulate materials, free liquids, and others. All these considerations tend to show a loss for the delivering party (Figure 9-2).

A dirty turbine meter can measure high or low, depending upon where in its range it is being operated. Gas quality has

Figure 9-2 The orifice plates shown here range from "clean" (upper left) to so dirty as to make measurement almost meaningless (lower right).

also suffered as maintenance for separators and dehydration plants has degraded. More carryover liquids are moving into pipelines. Add to this the difficulty in shutting in a poor quality gas source. With the pipeline "in the middle" between a purchaser that may be located remotely from the pipeline terminus and the producer interested primarily in continuing delivery, it is easy to understand why many meter stations have been shut in because of poor gas quality, only to be rapidly turned back on without addressing the original quality problems. The complexity of three-way negotiation among parties typically remote from each other makes the problem difficult (Figure 9-3a,b).

Experience has shown that a pipeline "lost and unaccounted for" report can be negatively impacted when this condition occurs. The solution is to improve the quality of the gas handled to the point that solids (iron oxides and sulfides and other pipeline dusts) and liquids (condensates and compressor oils) are reduced to within the contract limits. This minimizes effects on flow measurement. These same materials have a detrimental effect on the efficiency of the pipeline throughput. If allowed to go to extremes, they can cause damage to operating equipment such as regulators and compressors, in addition to the pipeline itself. Obviously such costly pipeline efficiency deteriorations and shutdowns are to be avoided.

Formation of the operation team concept or outsourcing has been less than effective in meeting typical maintenance

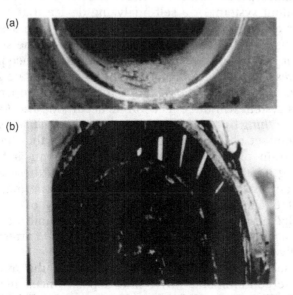

(a)

(b)

Figures 9-3 a, b Here are two more examples of dirty conditions found in the field: a meter tube and a turbine meter in pipeline use.

problems. A major problem is that the number of field personnel in the teams has been reduced from former levels. The teams have become generalist instead of specialist, and this has minimized their specific expertise in flow meter and system maintenance. The outsourcing contractor can report findings but is not a gas sales contract participant. The short-term results of reduced labor costs often ignore the long-term costs of sacrificed pipeline measurement and their effect on billings.

The routine work of today's field personnel is to emphasize the routine testing of the secondary element transducers and computers with less time allocated to the primary meter devices, which often just given cursory visual inspections. Little or no detailed meter tube removal and cleaning is scheduled. The resulting metering device may or may not meet the gas measurement standards requirements.

Requirements stated in contracts are based on the product whose quantities or contaminants are specified and limited. The contaminants are restricted to no material that will fall out in the pipeline after the gas has been delivered. It is these contract requirements that must be enforced to minimize maintenance. Consideration should be given to proper separation, filtration, drying, heating, or cooling and the costs of these operations to upgrade the gas to prepare it for measurement, rather than trying to keep meters clean in a dirty pipeline.

There is a desire simply to buy better metering to solve maintenance problems. This usually means a new meter, or a new readout system, or a self-analyzing device that sets off an alarm when it has a problem. While all of these are useful, dirty gas means that the user is normally exchanging one set of problems for another unless the gas quality maintenance problems are addressed. Buying a bigger computer to calculate incorrect data does not improve the data. No meter system operates in a dirty gas system without effects on its accuracy. *And there just is not such a thing as "clean" gas in the world's pipeline systems.*

Another set of ideas related to dirty meters that has started to appear in pipeline businesses says, "Give me a factor for correcting for the majority of errors caused by these problems and an analysis of the 'cost-value benefit' will be made versus the cost of cleaning up the gas." The fallacy of this is that knowledge of the correcting factors is lacking, and the database is poor or non-existent for all conditions, so the correcting data are highly questionable (Figure 9-4).

The second problem in this approach is the assumption that these contaminants arrive and deposit once, then do not change with time. Unfortunately, most flow and related contaminants along a pipeline vary from production to the final

Figure 9-4 A ball hone such as this one may be required to remove certain deposits in a meter tube.

customer. If wells load up and kick over, the operator hopes that separators will catch the fluids. Sometimes they do, often they do not. Likewise, dump valves on separators will hang up and fail to dump the fluids, which then end up in the pipeline. As compression and processing plants get older, there is more oil leakage, or more process upsets than is experienced with new equipment. Even if the contaminants are at very low values, as specified in contracts, over time they build up to significant total quantities.

For example, two molecules of steel pipe react with three molecules of oxygen to form ferric oxide. Whenever oxygen gets into a pipeline, chances are it will react with the pipe. Ferric oxide is found in all pipelines to some degree. It varies in color from red to black and is generally described as "line dust." As it is normally in the 1–3 micron size, it moves very easily with the gas flow, particularly as the flow rate changes.

To keep dusts from being formed, sources of oxygen—water, oxygen (air), and carbon dioxide—should be minimized in the pipeline. Hydrogen sulfide, although it does not contain oxygen, will react in the presence of oxygen to form several compounds of iron and sulfur. The quality requirement of natural gas was written around this knowledge.

Once formed, rust may be removed by pigging, filtering, or oil scrubbing. To minimize dust movement, low and stabilized velocities in the same direction are recommended. Variations in flow rates or reversal of flow will cause a "dust storm" as the rust is loosened. If it becomes a continuing problem, then oil fogging can be used to reduce dust movement. This, however, must be done with utmost care since the oil itself can cause trouble with the flow measurement.

The "bottom line" is that gas metering to the quality specifications of contracts is the simplest and best answer to minimize the maintenance problems.

Experience has shown that the best solution to a problem is to correct it at the source rather than to treat the symptoms after they appear in a pipeline system. Once started in a pipeline, they tend to be self-perpetuating and increase the problems in flow measurement over time.

In presentations at two meter schools, the problem of poor maintenance has been discussed in detail. In an article, "Problems in Offshore Measurement," presented in 1991 at the Acadiana School, examples of losses due to typical dirty measurement conditions were presented showing an estimated calculated loss of *$14.5 million per year.* A paper on the same subject at a later Acadiana School in 1996, entitled "The Value of Check Meter and Surveillance System," found the actual losses by using a check meter station in series with the billing system. The check meter showed losses of 0.42% of deliveries for a total of over $400,000 per year for five years, or slightly *over $2 million dollars of actual adjustments paid* because of the combination of primary and secondary devices found to be in error.

Many of the large adjustments were due to primary device errors that emphasized the need for these inspections because, as the second paper states, "all tubes will eventually foul" and "all stations are subject to problems."

Minimizing maintenance in an effort to reduce short-term costs has an economic impact on pipeline operation efficiency and metering in just the opposite way to that planned when the original downsizing was done.

Meter Tube Inspection and Cleaning

Good orifice meter measurement requires that new metering equipment is built and installed to meet the requirements of AGA-3 Part 2. As previously discussed, in addition to the requirements for new installations, meters must be maintained in like-new conditions to continue providing accurate measurement. This requires maintenance, since all pipelines contain liquids and solids that will dirty meter tubes and necessitate cleaning. Meter tubes for other types of meters should also be similarly inspected and cleaned.

Previous cleaning methods required line shutdown and removal of meter tubes. In recent years, with personnel and budget reductions in the operating and maintenance functions, it has become more and more difficult to get the proper meter tube care funded and accomplished. Yet dirty meter tubes are *one of the most common reasons why metering loses accuracy.*

New Cleaning Procedure

An improved method of cleaning meter tubes can circumvent the manpower/funding barriers. It involves using this equipment:

Fiber optic inspection device;

High pressure water pump;

Clean-water tank;

Dirty-water collection tank;

Various hoses used in pumping and collecting water;

Drain-collection open tank; and

Chemicals and/or ball hones needed for removing certain deposits.

Several types of hydraulic pump systems, tanks, hoses, hydraulic spray systems and chemicals, and ball hones are available from supply stores and service centers in most oil/gas areas where meter tube cleaning is needed. The equipment is usually rented but can be purchased.

The first step is to inspect the meter tube's internal condition with the fiber optic device. The meter tube must be blown down and removed from service. Insertion of the inspection device requires a tap that is 1/2 inch or larger in diameter. This may be a blow-off valve, a tap hole, or a fitting slot. Where no taps other than those required for T/P/dP transducers are allowed, a new tap may be installed in the meter station piping outside the distance specified by AGA-3.

The most critical parts of the meter tube to inspect are the 5 to 7 diameters immediately upstream of the meter and downstream of straightening vanes or flow-conditioner plates (if used). Examine these for any liquid or solid deposits. Contrary to common belief, *any amount* of deposit may cause measurements to be in error, usually low. Evaluation of the necessity for cleaning can be made after visual inspection, but cleaning is almost always the best choice. It is simply better to err by cleaning than to take a chance that the measurement may be several tenths of a percent low.

Cleaning also requires a means of inserting nozzles and hoses, typically a 1 inch or larger hole. This can be a 45° collar welded outside the dimensions specified by the applicable standard but inside shutoff valves. Such collars should be welded on the bottom side of the meter tube to allow drainage as well as hose insertion. The collar or collars should allow upstream and downstream access.

Manually feed the high pressure water hose and nozzle of the hydraulic system into the upstream tap. The downstream tap should have a collection container to collect dirty rinse water. Dirty water can be pumped from the container into a collection tank for proper disposal as an environmentally unsafe material.

The high pressure hose can be fed through straightening vane tubes 3/4 inch or larger in diameter. On small meter tubes—3 inches and less—the meter tube will have to be cleaned from both the upstream and downstream directions if flow-conditioning devices are present, or a smaller tube/nozzle system can be used.

The cleaning cycle should be repeated until the drain water runs clear. Check the actual results by repeating the optical inspection. If inspection verifies that internal cleaning has been effective, the drain plugs should be removed and all water drained from the fitting cavity. Proper choice of a drain tap depends on whether the fitting is upright (use both drains) or installed on the side (where only the bottom tap need be used for drainage). Standard purging and repressurizing procedures should be used to return the meter tube to service.

If a tube is not clean after high pressure washing, two other cleaning methods may be used; namely ball hone and/or chemical cleaning, depending on the composition of deposits. Chemical cleaning can be done in place, whereas ball-hone cleaning requires meter tube removal with a pipeline crew and lifting equipment (standard in the pipeline industry) (Figure 9-5).

Once a tube is cleaned chemically or honed, it may subsequently require only water cleaning for an extended time. Periodic inspection will establish how often cleaning should be scheduled and what type will be needed. Prices to rent inspection/cleaning equipment are lower than hiring out the work to be done by a pipeline or contract crew, but the cost and time associated must be added to rental costs for a meaningful comparison.

Figure 9-5 The orifice plate shown here in a gas measurement system clearly shows that a "river of liquid" was flowing in the line. Needless to say, gas measurement accuracy was far from accurate!

Summary

The need for clean meter tubes has been repeatedly demonstrated worldwide, yet it is one of the most common causes of measurement error. The procedures that have been outlined cost significantly less than the money represented by faulty measurement.

Effects of Liquids and Solids on Orifice Measurement

Experience with meter tubes with line dust (iron oxide, iron sulfide, dirt and glycol, or compressor oil) has been that the flow measurement is affected, causing the measurement to read low. The following three cases illustrate these effects.

Case 1

A pipeline meter station in the mid-continent area of the USA was examined in November 1999. This station had been running approximately 1.5 to 2% lower than a meter station in series with it for a "number of years." Many examinations and tests were run over the years to determine what caused the differences, with no solutions found. When the meter stations' setup was examined, this showed that there were approximately 50 yards between the stations with an oil separator halfway between them. The downstream station produced values 1.5% lower than the upstream. The meter tubes and orifice plates were removed; inspection revealed a light coating of oil (separator) with line dust. When this was mechanically removed, a repeated check between the two stations showed the second station to be only 0.2% lower than the first station—an improvement of over 1%.

Case 2

The inlet measurement at a gasoline stripping plant was questioned. When removed, the orifice plate and the meter tube were found with a 1/16″ coating of oil (compressor) and line dust. The accumulation was mechanically removed, and the meter increased its flow by 3%.

Case 3

A pipeline was losing approximately 0.6% of its throughput. This was a 1.2 billion cubic foot/day pipeline. The key sales meter stations were examined and found to have accumulations of compressor oil, scrubber oil, and line dust on the orifice plates and meter tubes. When this accumulation was removed the 0.6% loss became a slight gain.

Discussion

In each case, the meters were measuring from 1 to 3% lower than after cleaning. Based on these experiences and others, the meters would under-measure in the range of 1 to 3% as long as the coating of oil, line dust, and salt was present. The deposits change the relative roughness of the meter tube, and the velocity profile is elongated, which gives a low differential causing the lower volume rates. The orifice plate also changes its effective roughness, which affects the flow pattern.

As a result of the various cases previously discussed, a set of comparative tests were conducted using two orifice meters in series. A liquid was injected into the line downstream of the first meter and upstream of the second. This allowed the influence of the liquid on the downstream meter's readings to be compared to the readings of the non-influenced upstream meter. The results of these comparative tests were divided into three regimes.

The first flow regime had very low liquid content. The liquid content was just enough to produce a thin coating on all of the orifice meter's internal surfaces. Research performed with low injection rates has shown that flow is under-measured in such circumstances. The under-measurement is consistent and in some cases can be over 1%.

The second flow regime used a higher liquid content than the first regime. The liquid content was high enough for a thicker layer of fluid to form at the bottom of the pipe. This thicker layer of fluid could become redistributed around the pipe surface if fluid velocities increased. Pooling of the liquid in front of the orifice plate could also occur. In this flow regime there is a lot of variability in the behavior of the liquid, and as a result the flow measurement may not be affected or there may be substantial under-measurement of flow. The boundaries of regime 2 can extend from 0.0% or slightly positive flow measurement error to under-measurement of as much as 1%.

The third flow regime has enough liquid in the pipe to create a permanent thick layer of fluid on the bottom of the pipe. The surface of the liquid has a fairly constant roughness. The liquid is dammed up in front of the orifice plate and is drawn through the orifice bore at a steady rate. Orifice plates experiencing liquid loads this high consistently over-measure flow. The amount of over-measurement appears to be a function of the flowing gas pressure and extends from 0.0%, or slightly positive flow measurement error, to as much as 6% over-measurement.

Clearly the issue of free liquid's influence on orifice metering is variable and results in unacceptable accuracy. Testing of other meter devices show similar results but has not been as definitive.

Flow Computers Require Maintenance

Electronic flow meters/computers (EFMs) are being used increasingly on gas pipelines to meet FERC Rule 636 requirements and the need for online, real-time measurement.

Information from the computers fed to bulletin boards also is used for controlling pipe line inputs and outputs to balance the needs of customers against supplier inputs. The need to quickly determine flow volumes has diminished the use of recording charts to a secondary basis on smaller pipe line input and output volumes.

As EFM capabilities grow and prices drop, their use continues to increase. They generally are selected without question over charts when economics are favorable. Many manufacturers produce a wide variety of models to meet varied customer needs. *All computers require maintenance.*

EFMs were originally sold on the basis that they were the "way to go" or "the modern approach." They were designed to be fast electronic calculators that followed the procedures used with charts. They have proved their usefulness in terms of reliability, flexibility, data availability, and speed. Not everything about them, however, is completely rosy. For instance, where does the output come from? The first concern of an astute user should be to question the acceptance of digital output values that involve little or no checking of input data.

Charts by their nature provided an initial source to review data. Chart auditing was the first step in chart processing. If a chart's recording was unusual, an experienced auditor would notice it immediately and question the field operator to explain or reconfirm recordings prior to chart calculation. In the vast majority of situations today, there is no comparable input data review for EFMs.

The EFM output checking that does take place is usually a review of digitized flow values. Unless an obvious discrepancy appears, there is no questioning of the validity of inputs from which volumes are computed. This "after-the-fact" auditing tends to obscure problems that would have been noticed with charts prior to volume calculation.

Chapter 21, "Flow Measurement Using Electronic Metering Systems," in the *API Manual of Petroleum Measurement Standards* (MPMS Chapter 21) lists the information required: daily and hourly quantities for time, flow rate, differential pressure (for orifice use), static pressure, and specific gravity (relative density). In normal operations, time, differential pressure, static pressure, specific gravity and heat content are generally not reviewed unless volumes or totalized heat are questioned. The audit trail listed in MPMS Chapter 21 lists these records to be reviewed:

1. Transaction records;
2. Configuration logs;
3. Event logs;
4. Corrected transaction reports;
5. Field test reports; and
6. Minimum data-retention periods.

These data will provide sufficient information to determine reasonable adjustments when:

- The system stops functioning;
- The system is detected to be out of accuracy limits from maintenance tests; and
- Measurement parameters are incorrectly recorded.

The problem, as stated above, is that the hourly data, even though properly recorded, usually are not completely audited unless an obvious problem appears with calculated volumes. In other words, flow measurement quality control has been greatly de-emphasized. For this reason, several pipelines have experienced ungainly increases in their "lost and unaccounted for" gas.

Some typical problems that were detected with chart processing, but that remain routinely undetected when EFMs are used, include low differential pressures, improper temperature measurement, loss of static pressure regulator controls, improper specific gravities, and improper heat content. The computer "spits out" flow volumes and total delivered heat, and the user accepts the numbers without question.

The basic problem, therefore, is that responsibility for additional auditing is all too often not defined clearly—if at all. (See "Who's Watching the Cash Register" in the September 1996 *Pipe Line and Gas Industry*, page 94.) Better auditing is usually triggered only when discrepancies show up in "lost and unaccounted for" reports. There is no telling how many such erroneous reports have gone into the system undetected.

With economic pressures forcing field personnel reductions, it is valuable to schedule maintenance on an as-needed basis. When volumes and total heat values are properly audited, they become a good source for such maintenance scheduling. The consequent efficient use of field personnel pays off in appropriate testing, better maintenance, improved flow measurement, and ultimately, improved lost and unaccounted for balances.

Still, it is important to remain alert. The "garbage in, garbage out" rule is just as applicable in flow measurement as it is in any other calculation. Beware of the hidden trap with digital techniques. Just as printing a sentence does not necessarily make it a fact, printing out digital values does not make them necessarily valid. And the responsibility for thorough and regular data auditing needs to be clearly established.

Effects on other Meters

Ultrasonic meters are also affected by deposits. The influence of fouling on the accuracy of the ultrasonic gas flow measurement is partly generic and partly dependent on the design of the meter. The first relates to the change in the cross-sectional area; the second depends on the transducer design and frequencies used as well as the path configuration.

All flow meters—including orifice, turbine, and ultrasonic— are sensitive to the stability of their cross-sectional area. When fouling is present, the uncertainty of the cross-sectional area has by far the largest influence on measurement uncertainty. Being a "squared relationship," an 0.1% change in the diameter leads to an 0.2% measurement error.

In dirty gas applications, small ultrasonic meters are more sensitive than large ones, but for all meters it is essential to determine whether and possibly how much fouling is present. Here the comparison of single- and multi-reflection paths in a meter offers better diagnostic analysis over straight path designs (but not necessarily better overall operation and accuracy).

Fouling on ultrasonic bore walls and transducer surfaces can produce three sources of error; measuring effects from them can each also help diagnose possible problems:

- Signal attenuation;
- Decreased path lengths;
- Changes in flow profile and possibly in swirl angle; and
- Changes in signal strength, which can be caused by fouling on the transducer head, or for reflective designs, a change in the ultrasonic reflection coefficient.

Operation over a wide range of pressures (up to approximately 9,000 psi/600 bar) and varying gas compositions—especially offshore in situations in which pipeline dirt and oil are coming down the pipe—requires ultrasonic transducers to cope with a wide range of acoustic impedance.

Installation effects can also influence ultrasonic meter uncertainties. Upstream piping (elbows, tees, abrupt changes, various other swirl generators) that causes the fluid flowing through the line to become non-uniform can cause a calibration shift and/or distortion in the flow signature (velocity profile and swirl angle). A non-representative sample is then measured by the ultrasonic path(s) to yield measurement uncertainty. Multipath meters help read such flow distortions appropriately, with some designs doing a better job than others for various reasons. Coriolis meters claim that distortions in flow profiles do not affect meter accuracy except in very extreme situations.

Meter manufacturers offer options to alarm and/or correct for these conditions. It remains to be seen how successful these are. In any event, the safest procedure is to make sure the ultrasonic meter is kept clean.

General Maintenance of Liquid Meters

Most oil and gas industry meters are proven on a schedule set by the contract, company policies, or government requirements. In every case, these provings are the basis for initiating maintenance and/or proof of meter accuracy. All proving should be carefully run with witnesses from the contracting parties being present. Both parties to the proving should be trained and knowledgeable about proving requirements as specified in the API standards. The standards are the basis for contractual requirements for the exchange of the liquids. The provings must be run correctly to establish meter accuracy and thus control the correct billing for custody exchange. The provings become a legal document in case of a meter settlement lawsuit. (Such provings for gas measurement are more often the exception rather than the rule. The throughput test is the basis for accurate liquid metering.)

The proof of accurate metering is the balance between liquids into and out of the system adjusted for inventory changes of liquid in the system. Most often the balances are quite close on a well-measured system, but a running balance must be kept that will indicate any changes occurring in the system. These balances should be kept on a given time period such as day, week, month, or year. Shorter-term balances will show more excursions but are useful for spotting troubles and initiating maintenance before the system accuracy is lost beyond some agreed upon percentage of total deliveries. A typical target balance is between 0.1 and 0.25%. To a degree, the value chosen reflects the complexity of the system and control of maintenance practices. Pipelines use loss allowances to off-set the financial loss due to the uncertainties of the liquid pipeline measurement. In a system that is handling hundreds of millions of dollars worth of liquids, an allowance of only 0.25% (a common value used) can result in millions of dollars of lost revenue; vigorous effort should be invested to keep measurements correct so the lost revenue will be as close to zero as possible.

Tracking system balance reports are usually reported on control charts that display visually the gain and loss of a system. They determine the stability of the system and show any trends or step changes in the balances. They are the basis for initiating studies to improve performance. This performance evaluation, a detailed review of proving and maintenance reports to spot any troubled meter stations, should include a study of:

Personnel: how properly trained and committed;
Procedures: how well API requirements are met;
Facilities: built how close to API standards;
Equipment: operating in proper range;
Calibration: done correctly and on proper schedule;
Piping: no leaks or improperly operating valves;
Computer/calculation: correct programs and data correctly used; and
Security: facilities secure to unauthorized personnel.

A completion of the study should point out any troubled stations and what needs to be changed in their operation and maintenance to improve their accuracy.

Specific Liquid Maintenance Problems

In addition to the straightforward maintenance problems of metering systems covered so far, there are unusual problems that can be experienced in special instances.

Stabilized crude oil is crude stored at atmospheric pressure to allow light-end components to be reduced over time. Stabilizing crude oil minimizes the possibility of two-phase flow resulting from light ends breaking out in the metering system. Pressure and temperature correction factors in the ASTM/API tables for reducing liquid measurement from flowing to standard pressures and temperatures are based on stabilized crude. Use of these factors on unstabilized crude can produce a significant error. Thus, trying to measure two-phase flow with conventional meters and using correction factors incorrectly should be avoided.

Waxy crude, crude oil with high percentages of wax, can create problems if the "wax point" temperature is reached. This is a particular problem when the temperature is variable in a system; for example, a North Sea system normally operated 30°F above the wax point, because of the well production temperature. A multiple turbine meter prover system operated well for a period of time, but suddenly the meter factor tests gave highly erratic results for all the turbines.

The turbines were removed for inspection, but nothing untoward was found. After repeated provings, the turbine metering was resumed. Again the values became erratic. The operator removed the meters for inspection. This time there was a light coating of wax on all meter parts in contact with the crude. This meter system was located at a central platform, below which one million barrels of oil were stored in the cold North Sea. This storage was used as an emergency reserve when pipeline operating problems occurred and only occasionally used for delivery. The length of time that the crude stayed in storage varied. Proving became erratic after crude was delivered from the storage after it had been held for some time. When warm crude was returned to the meter, the deposited wax was melted. This caused the on-again, off-again erratic meter factors.

The problem was solved by the meter specialist knowing that meter and prover systems would not act erratically unless something had occurred. In this case, normal maintenance found nothing, but the combination of knowing performance characteristics and personnel persistence paid off by finding the problem, even though it was transient as the wax deposits came and went.

Another unusual proving and turbine meter installation with confusing results occurred at two multiple turbine stations that were located on either side of a prover. One set of meters proved as expected, but provings for the other set (using the same prover) would not repeat. To study the problem, a meter from one station was exchanged for a meter from the other station. The provings showed that the meter that was good became erratic, and the bad meter became good in its new location. Both stations were mirror images of each other and were built to the API standards by the same manufacturer. Once again, the meter specialist knew that there had to be some difference that was related to the specific locations.

On further study, it was noted that the two stations fed from one pipeline with one elbowed off to the right and the other elbowed to the left. So the inlet piping was 180° different as it entered the station headers. In reviewing this, one station had an upstream piping that caused the flow to swirl clockwise and the other counterclockwise. Both stations had proper lengths and straightening vanes. The solution suggested was to put in doubled length vanes. When this was done, the provings in the bad station became repeatable.

The purpose of these three examples is to point out that the guidance for maintenance in the API standards is a good base, but intelligent analysis of the meter system may also be required to get the best flow measurement.

MEASUREMENT AND METERS

After identifying—and addressing the care of—all effects of fluids and the flow signature for a specific application, meter selection and application can be attacked. Several general considerations as well as specific characteristics of individual meters will be reviewed.

Meter Characteristics

Comparing Meters

Characteristics to be considered when evaluating a meter include: achievable uncertainty, comparative cost, use acceptance and specific use, repeatability, maintenance costs, operating cost, few or no moving parts, ruggedness, service life, rangeability, style to meet fluid property problems, pressure and temperature ratings required, ease of installation and removal, power required, pressure loss caused by meter (running and stopped), and how well calibration can be proved.

No single meter will have all of the characteristics desired, but candidates can be evaluated by going through such a list for each meter under consideration and then deciding which of the factors are of prime importance for the particular flow measurement problem. A procedure something like the following may be helpful:

1. List, for each candidate meter, the characteristics of importance.
2. Define *how important* each is by assigning a weighting factor (such as from 10 for very important to 0 for no importance).
3. Assign a similar rating number to show how well the meter will *perform* to meet specific needs of the application.
4. Multiply the weighting factor by the performance factor, and add all the totals.

Comparing the totals for various meters will provide a rank ordering. (This same concept can be applied to the total

Fluid Flow Measurement. ISBN: 978-0-12-409524-3

159

metering system problems including considerations for flow, fluid, installation, maintenance, etc.)

There are no industry-required standards or testing facilities where meter tolerances can be validated and a ranking of meters published. Even if such a facility were available, it would be impossible to make a test expressing all possible uses of a meter. A simple case in point: A test on a gas may not relate to a liquid metering problem or vice versa, even with consideration of the Reynolds number effect. Therefore, specific information may not be available, but knowledge *is* available in the industry to give *guidance* on the probabilities of success for a given meter in a given application.

Specific meters prove useful with specific applications and/ or fluids—or they quietly disappear from the marketplace. The flow measurement fraternity is a relatively conservative group. New meters based on a new principle have a long road to acceptance under the best of circumstances. The history of every meter in use today has followed this road. A meter must provide acceptable performance/cost ratios, considering initial cost plus the other required investments in proper installation, operation, and maintenance, or it will not complete the trip down the acceptance road.

Types of Meters

There are many ways that types of meters can be listed; they are listed here by their individual names under general categories, since there are significant differences between the meters in the same category (i.e., differential head meters of the orifice, flow nozzle, Venturi, and Pitot tube types; this is true for the linear meter). They are listed in random order, and there is no significance to the order.

The following charts show major flow meters and their salient characteristics. These are by no means definitive since they list only general categories and performances. Most manufacturers claim performance better than tabulated here—but values in the tables are more typical of *actual field conditions* for properly installed and maintained meters. For specific applications, the comments may or may not apply. Once again, reference should be made to relevant industry standards and data, manufacturers' information, and the experience of others to evaluate flow metering for a given application. But these tables provide a starting point (Figure 10-1, Tables 10-1, 10-2).

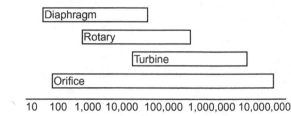

10   100   1,000 10,000 100,000 1,000,000 10,000,000
cfh of natural gas at base conditions

Figure 10-1 Comparison of flow ranges for various gas flow meters.

Flow Meter Specification Terms

Meter accuracy is a much-abused term. Manufacturers often default to it for "selling shorthand," but those specifying and buying meters to measure flow should be aware of several caveats.

Consideration of meter characteristics shows clearly that simply relying on a manufacturer's statement of accuracy is indeed an incomplete and inadequate way to evaluate and compare meters. And, as stated, proper determination and application of accuracy, rangeability, linearity, repeatability, and hysteresis data are basic but still only part of the job in achieving the best flow measurement; operation and maintenance must also be considered.

Meter factor is a dimensionless correction that mathematically modifies a meter's indication to a corrected true reading based on knowledge of the flow and flowing conditions. Corrected readings may be manually calculated periodically or the meter factor automatically applied continuously. It is normally determined from a throughput test, covering the range of flows to be measured, based on a master meter or a prover.

In the first place, no meter is absolutely accurate!

There are no absolute standards of gas or liquid against which to compare a meter reading to see how closely that reading compares with what is *actually* flowing through the meter. Furthermore, any statement of accuracy must include not only the best possible estimate of how accurate the measurement is but also over what flow range the estimate applies. The more diligent manufacturers usually supply such detailed information. But that is only part of the problem in determining meter accuracy (Figures 10-2, 10-3).

As previously noted, it is equally vital to know and take into account the type of fluid being measured, the conditions under which the meter will be used (including fluid condition), how it

Table 10-1 Characteristics of Meters for Gas Measurement

Meter Type	Accuracy of Sensor[1] Proved (±)	Unproved (±)	Rangeability (Range for Stated Accuracy)	Reynolds Number (Minimum)	API Upstream Piping (X-Diam.)	Pressure Limit (psig)	Temp. Limit (°F)	Notes[1] - System Accuracy Depends on System Design, Quality of Maintenance, etc.
HEAD METERS *(Flow Proportional to $[\Delta P]^{0.5}$)*								
Orifice	0.25	1.50	3:1	4,000	See Ch. 11	ML	ML	Square-edged, concentric
Flow nozzle	0.50	1.50	3:1	10,000	10 – 40	ML	ML	ASME
Venturi	0.50	1.50	3:1	7,500	5 – 30	ML	ML	
Elbows	0.50	3.00	3:1	10,000	10 – 40	ML	ML	Prepared as a meter
LINEAR METERS *(Flow Proportional to Flow Velocity)*								
Non-intrusive ("Open-pipe" bore)								
Coriolis	0.5	1.50	5:1 to 25:1	3,000	None	ML	ML	
Ultrasonic	2.00	5.00	10:1	5,000	10	To ANSI 2500	–300 to 200	
Doppler								
Transit time	0.50	1.00	10:1[1]	3,000	10	To ANSI 2500	–300 to 200	
Intrusive (Sensing element intrudes into bore)								
Multiphase	5 – 10	5 – 10						
Turbine, gas	0.25	0.3	10:1 to 140:1	4,000	10	1,400	150	
Vortex shedding	0.5	2.00	10:1 to 50:1	4,000	5 – 40	2,500	400	

[1] Extended low-flow meters rangeability to 50:1; ML = Material limit

Table 10-2 Characteristics of Meters for Liquid Measurement

Meter Type	Accuracy of Sensor Proved (±)	Unproved (±)	Rangeability (Range for Stated Accuracy)	Reynolds Number (Minimum)	Piping Required (Diameter, inches)	Pressure Limit (psig)	Temp. Limit (°F)	Notes[1] System Accuracy Depends on System Design, Quality of Maintenance, etc.
HEAD METERS (Flow Proportional to $[\Delta P]^{0.5}$)								
Orifice	0.25	1.50	3:1	4,000	See Ch. 11	ML	ML	Square-edged, concentric
Flow nozzle	0.50	1.50	3:1	10,000	10 – 40	ML	ML	ASME
Venturi	0.50	1.50	3:1	7,500	5 – 30	ML	ML	
Elbows	0.50	3.00	3:1	10,000	10 – 40	ML	ML	Prepared as a meter
LINEAR METERS (Flow Proportional to Flow Velocity)								
Non-intrusive ("Open-pipe" bore)								
Coriolis	0.25	1.00	5:1 to 25:1	3,000	None	ML	ML	
Ultrasonic Doppler	2.00	5.00	10:1		10	To ANS 2500		
Transit time	0.50	1.00	10:1[1]		10	To ANS 2500	−4 to 185	
Intrusive (Sensing element intrudes into bore)								
Multiphase	5 – 10	5 – 10			None			
Positive displcm't	0.25	1.00	20 – 1		None			
Turbine, liquid	0.10	0.50	10 – 1	No limit	10	1400	150	
Vortex shedding	0.5	2.00	10:1 to 30:1	10,000	5 – 40	2500	400	

[1]Extended low-flow meters rangeability to 50:1; ML = Material limit

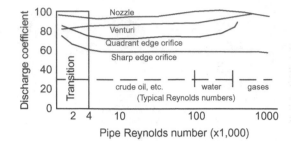

Figure 10-2 Shown here are typical effects on discharge coefficient for various meters and bias error versus Reynolds number in measuring typical fluids.

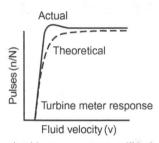

Figure 10-3 Many factors besides accuracy specifications are important for a flow meter performance curve. For example, meter response speed may depart from theory.

will be installed, how it will be operated, and what level of maintenance will be provided. Otherwise, accuracy statements are meaningless in terms of the values actually obtainable.

Rangeability expresses the flow range over which a meter operates while meeting a stated accuracy tolerance. It is often stated as "turndown," maximum flow divided by minimum flow over the range. For example, a meter with maximum flow (100%) of 100 gallons per minute and minimum flow (within a stated tolerance such as ±0.5%) of 10 gpm has a 10-to-1 rangeability or turndown of 10. It will be accurate ±0.5% from 10 to 100 gpm.

The meter may provide a tighter tolerance over a more limited range—say 10 to 1 within ±0.5% of actual flow as above, but 3 to 1 within the ±0.25% accuracy range. This means the user can select the tighter tolerance of ±0.25% for a range 33 to 100% flow (33 to 100 gpm).

Linearity defines how close to a specific accuracy the meter registers over a stated flow range. Its proof curve will

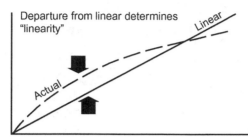

Figure 10-4 Linearity describes how well a meter tracks a straight line proof curve.

approximate a straight line. It may be significantly inaccurate but quite linear (Figure 10-4).

It is important to point out that this characteristic was much more important prior to the widespread use of computers and electronic signal-conditioning equipment. With a computer correction device, it is possible to characterize a non-linear meter output curve *provided the meter output is repeatable.* Such curve characterization allows a closer fit of the readout system—even for a linear meter—to minimize calibration errors. The same procedure is used on "smart" transducers to minimize any non-linearity the transducer may exhibit as a result of temperature and/or pressure effects.

Repeatability means just what it says: how nearly the same reading a meter will provide for a series of measurements of the same quantity and procedure carried out by the same operator and the same equipment in the same location over a short period of time and given flow condition. As with linearity, it may be more important to always get the same reading for specific flow rates than it is that those readings are extremely accurate. Flow control is an example in which repeatability is more important than accuracy.

Hysteresis is closely related to repeatability. It describes what happens to meter output as a given flow rate is approached from larger and smaller rates. For example, suppose a flow rate of 80 gpm is increased to 100 gpm, and a meter then registers 99 gpm. Now the flow rate increases to 120 gpm and returns again to 100 gpm; the meter registers 101 gpm. Its hysteresis is ±1 gpm, and the dead band is 2 gpm at a 100 gpm flow rate.

DIFFERENTIAL (HEAD) METERS

For several centuries, the basic concepts of differential meters have been known: (a) flow rate is equal to velocity times the device area, and (b) flow varies with the square root of the head or pressure drop across it. Likewise, the equations of continuity were well known.

The first two systems designed on these basic concepts were the Pitot (1732) and the Venturi (1797). The flow nozzle was used in the late 1800s, and the orifice appeared in commercial use in the early 1900s. In each of these cases, the original investigators set their own requirements of design configuration, calculation, installation, and operation of their units. Continued research on and use of the various head devices since this early work has resulted in updated standards for their construction, installation, operation, and maintenance. Research continues today, reflecting the continued interest in the differential meter as a useful flow metering device (Figure 11-1).

The head device consists of a primary element that restricts the flowing stream, resulting in a pressure difference (pressure drop) across the primary element. This differential pressure relates to the flow rate through the restriction by application of Bernoulli's equation. However, use of the equation assumes the physical area through the restriction to be the flow area through the restriction—and this is not the case with differential head meters where the physical area through the meter is always greater than the effective flow area. Bernoulli's equation must therefore be adjusted to account for the difference in areas. The adjustment, called the "coefficient of discharge," is a function of the meter's physical area, differential pressure tap location, and Reynolds number. It can be calculated from equations developed from empirical data on each differential head meter. It is very important to remember that to take advantage of a head meter's calculable performance, based on its mechanical construction, these specifications must be known and followed. Any deviation from these requirements necessitates a flow calibration to determine specific relationships of pressure drop to flow rate.

Fluid Flow Measurement. ISBN: 978-0-12-409524-3

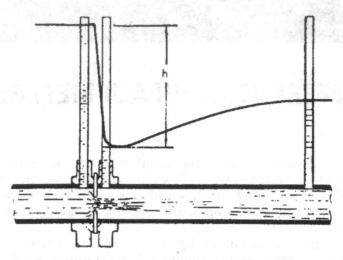

Figure 11-1 Typical orifice flow pattern and pressure differential across the orifice.

The related secondary devices consist of a differential pressure measuring unit with connecting piping and other measuring units required to define the flowing variables of the fluid, such as pressure, temperature, and composition. The pressure and differential pressure transducer is often combined into a single unit.

The advantages of a head meter can generally be listed as follows:

1. Simple;
2. Inexpensive;
3. Calibration inferred from mechanical construction;
4. Available in most sizes;
5. Easy maintenance;
6. Rugged;
7. Widely used and accepted;
8. Does not require power; and
9. Can be built with special materials.

Specific head meters may not have all of the above advantages, but these are the general considerations for choosing a head meter. For example, the Venturi and the Pitot meters have relatively low permanent pressure losses. Others, such as the orifice, are adaptable to capacity changes by changing plates, provided the capacity change is not rapid with time. The Venturi and the nozzle will handle dirty fluids but they are not easily removed for maintenance (Figure 11-2).

Head devices have few application limitations and are used for most types of fluid flow measurement. They have been particularly applied in the measurement of water, natural gas,

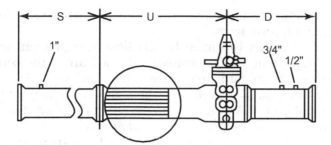

Figure 11-2 Typical orifice meter built to AGA-3.

hydrocarbons of all kinds, chemicals, etc. It is easier to list the fluids where they are not used, such as viscous liquids, slurries, pulsating, multiphase, and non-Newtonian. Even these limits can be handled with special care under some circumstances.

In summary, the head devices cover a large category of flow meters in service today. Their wide acceptance and use is based on successful applications and service over many years.

Orifice Meter

Construction of the orifice meter is extremely important, since its mechanical construction defines its calibration under the various applicable standards. One of the most complete standards written on the orifice meter is the ANSI/API, Chapter 14, section 3, AGA Report No. 3, "Natural Gas Fluids Measurement," more commonly called "AGA-3."

AGA-3 is the bible of the orifice meter for gas meters. It represents the compilation of many different tests covering about 90 years with the latest agreed-upon knowledge of construction and installation, method of computing flow, orifice meter tables for natural gas, nomenclature and physical constants, and appendices. It is subject to periodic review and is updated as new knowledge is gained. It represents the most widely used standard for high pressure natural gas measurement and is successfully used commercially to measure the majority of the natural gas exchanged in the world.

Meter Design Changed

The following outline shows how an established standard can be changed because of new test results. All standards are living documents and must be continually followed to implement the improvements.

Detailed review is a must for the flow measurement practitioner. Depending on the words used, the part of the contract on the use of standards may state "The standard to be used is the latest edition of AGA-3," or "The standard to be used is the latest edition of AGA-3 as mutually agreed upon," or "The standard to be used is the [date] edition of AGA-3."

Because of these variations, the use of the standard AGA-3 Part 2 is not universally used or understood. The basic contention of the standards is that these requirements must be met on all new installations or when an old station is reinstalled. The mechanical requirements of the meter tube description, fitting requirements, flow conditions, orifice plate requirements, and installation are the major sources of the changes listed below.

Meter Tube and Fitting

Inside surface (smoother);
Diameter (tighter tolerances);
Roundness (tighter tolerances);
Orifice plate sealing or sealing devices, recesses and protrusions (tighter tolerances);
Flanges (see all of the above);
Welded attachment to pipe (see first three above);
Inspection (to prove compliance);
Bypass (leakage); and
Pressure taps (tighter tolerances).

Orifice Plate

Faces;
Bore edge (more defined);
Bore diameter;
Bore thickness (redefined);
Plate thickness; and
Bevel (redefined).

Flow Conditioner

19 tube bundle (further limitations);
Other conditioners (covered by defining tests); and
Performance criteria (tests for new flow conditioners).

Installation Requirements

Fluid dynamic conditions (further defined);
Plate eccentricity (tighter tolerances);
Plate perpendicularity (defined);
Meter tube: length (changed); and
Meter tube: requirements for flow conditioners (changed).

It is important that those knowledgeable in the practice of the standards recognize that the requirements have changed. To be in compliance with the new requirements, all of the changes must be made in new installations to obtain the uncertainty limits of the standard. Failure to do so may compromise the stations' volumes. As time goes on it is expected that upgrading of the orifice standard will be continued. To take advantage of these new improvements, developments must be followed continually.

AGA-3 is also used to define an orifice installation for liquid flows. Its usefulness relates directly to understanding its contents. An introduction to the standard is given here, with the hope that this will form a basis for further study to properly complete the job of preparing system designers and users to measure with an orifice meter. The standard is based on a very large number of tests and the experiences of flow measurement practitioners; to use the report effectively, conditions found to be significant must be reproduced.

The basic premise of the standard is to make it unnecessary to calibrate each individual meter, but be able to predict and control its measurement accuracy by controlling how the meter is made and installed. This has been found possible to a tolerance acceptable for commercial measurement (custody transfer) for head devices.

The orifice is the most important and most widely used of the primary element head devices. It comes in many configurations for special applications, but by far the most common is the sharp-edged, concentric orifice plate using flange taps (located one inch from the upstream face of the plate and one inch from the downstream face of the orifice plate).

Reynolds numbers and beta ratios are the keys to the accuracy limits of plates. With a beta of roughly 0.10 to 0.75, the percent of uncertainty is 0.5% or less. However, at Reynolds numbers below 100,000 with large beta ratio (0.75), uncertainty is increased as shown in Table 11.1.

Uncertainty becomes so large—several percent—for Reynolds numbers below 4,000 that accurate measurement is no longer possible. The lowest Reynolds numbers typically occur on smaller runs with smaller plates on fluids more viscous than water.

Table 11-1 Concentric sharp edged orifice plate coefficient of discharge uncertainty versus Reynolds number

Reynolds Number	Uncertainty (%)
100,000	0.58
50,000	0.59
10,000	0.70
5,000	0.84
4,000	not recommended

An orifice used on liquid is limited by the lower pipe Reynolds numbers listed above, which are often more important because of the higher viscosities of liquids. The reason is the non-linearity of the flow coefficient with slight changes in the Reynolds number; at these low ranges, the changes in flow coefficient cause significant errors in flow rate measurement unless coefficient, volume, and Reynolds number calculations are iterated continually. For accurate measurement, rangeability of an orifice with a 100 inch recording device is usually limited to 3 to 1 for a single installation. However, multiple differential readout devices or multiple tubes are commonly used to extend the orifice rangeability.

The major orifice use is for general purpose, non-viscous flows where low cost is important. Compared to some other meters, it has a relatively high permanent pressure drop for a given flow rate, which may limit its use if pumping costs are a major consideration. Maintenance of the primary device consists of a periodic inspection and cleaning of the plate and the meter run, since inaccuracies can occur if initial equipment condition is not maintained. The primary advantages of the orifice use are its wide acceptance, simplicity, the large number of trained operating personnel available to maintain it, and the large amount of industry research data available on it.

The orifice standard is: ANSI/API 14.3/AGA-3 (latest edition). Its usefulness depends directly on how well the information in the standard is applied. Subjects covered by the standard include:

Part 1, General Equations and Uncertainty Guidelines
Part 2, Specification and Installation Requirements

Part 3, Natural Gas Applications

Part 4, Background, Developments, and Implementation Procedure and Subroutine Documentation for Empirical Flange-Tapped Discharge Coefficient Equation

Part 1 of the standard contains the basic flow equations for all fluids to be used to calculate flow rate through a concentric, square-edged, flange-tapped orifice meter. It provides an explanation of the terms used and methods for determining fluid properties. A definition of the uncertainty and guidelines for calculating possible errors in flow of fluids are given.

Part 2 of the standard was revised in April 2000 and gives specification and installation requirements for the manufacture and use of orifice plates, meter runs (upstream and downstream) orifice plate holders, pressure taps, thermometer wells, and flow conditioners. These specifications and requirements must be followed in detail to use the uncertainty calculations of Part 1. However, the standard recommends use of updated calculation methods, even if field equipment has not been updated to meet the new physical requirements. Variations from the requirements on initial installation, as well as during use of the station, will create the potential for error in flow measurement. In-place testing is required to determine what calibration factors should be applied, since the error estimates of the standard do not apply to non-standard metering situations.

Part 3 is specific to the use of the standard for natural gas measurement as practiced in North America.

Part 4 discusses the background, development, and limitations of the coefficient of discharge data. This material is of academic interest but not required to apply the data. However, the second section of Part 4 is important because example calculations detail the procedures to implement the data. The purpose of this section is to minimize accounting differences obtained by various computers and/or programs in using the data.

The latest version of the new document is different from previous standards in some details, such as the equations used and the requirements for manufacturing and installing an orifice. A grandfather clause allows stations under previous standards to continue to be used. The new standard contains the latest, most defensible data, and it reduces the uncertainty of measurement over a wider range of application of the orifice for flow measurement.

Other standards that cover the use of orifices include ISO 5167 and ASME MFC-3M. Since both trace back to the same basic databases, there are similarities between the documents. However, there are also differences that must be understood, since an orifice built to one standard may not meet the other standards.

The measurement fraternity is attempting to resolve these differences. In the meantime, if a specific standard must be met, then it must be specified in ordering and operating equipment.

Orifice Meter Description

An orifice meter consists of an orifice plate, a holding device, upstream downstream meter piping, and pressure taps. By far the most critical part of the meter is the orifice plate, particularly the widely used square-edged concentric plate, whose construction requirements are well documented in standards such as AGA-3 and ISO 5167-1. These standards define the plate's edge, flatness, thickness—with bevel details, if required—and bore limitations.

The most common holding system is a pair of orifice flanges. However, for more precise measurement, various fittings are used. These simplify plate installation/removal for changing flow ranges and for easy inspection. In every case, the orifice must be installed concentric with the pipe within limits stated by the standard.

An orifice plate installed without specified upstream and downstream lengths of pipe controlled to close tolerances and/or without properly made pressure taps (usually flange) is not a "legitimate" flow meter; it must have specific tests run to determine its calibration. Since this is not economical, almost all orifice systems are built to meet the standard(s). This allows calculations to be made with specified tolerances. Control over orifice metering accuracy derives directly from data in the standard, which must be followed without exception.

Sizing

Sizing an orifice depends on the flow measurement task to be done. For example, a simple design would be a single meter tube with a single orifice plate in the mid beta ratio range, which would be sufficient for a fairly constant flow. However, if the flow grows over time, then sizing should allow for this growth by using a 0.20 beta plate. If flow is likely to decrease with time, then a large beta (such as 0.60) should be used. This way the meter tube size will not have to be changed in a short time period.

If continued growth is anticipated, an economical design would be to size the lengths and valves for the meter tube size slightly larger (i.e., if a 6 inch tube is chosen, sizing the meter tube length for an 8 inch tube, which could then replace the 6 inch without piping and header changes).

On non-custody transfer metering where utmost accuracy is not of prime importance, design betas can be changed to 0.15 and 0.70. If flow changes take place in a short time period, consideration should be given to several ranges for differential devices tied to one orifice or the use of multiple meter tubes switched in and out of service automatically. The differential chosen to size an orifice should not be full scale, but a reduced differential to allow for some inevitable variations. The amount of this reduction might be 10 to 20% of range depending on the likelihood of this much flow variation occurring.

Sizing of orifice stations is relatively simple for fluids with steady flow rates. Since a single orifice with a single readout system is limited to a flow range of roughly 3 to 1, for most accurate measurement, knowledge of the flow ranges is required to properly size the orifice to prevent over or under ranging. The controlling design flow rate should be the "normal flow" since the majority of the flow will, by definition, be at this rate. If the limits are beyond the 3-to-1 range, multiple orifice or readouts or possibly another meter should be chosen. There are many calculator and computer programs available in the industry to assist in this sizing. Likewise, manufacturers offer sizing calculations as part of their service of manufacturing plates.

Equations

The basic orifice meter mass flow equation in the common US system units (IP) is as follows:

$$q_m = C_d \, E_v Y(\pi/4)d^2(2g_c \, \rho_{t,p}\Delta P)^{0.5} \qquad (11.1)$$

where:

q_m = mass flow rate (lbm/sec);

C_d = orifice plate coefficient of discharge (dimensionless);

E_v = velocity of approach factor (dimensionless);

Y = expansion factor (dimensionless);

π = universal constant (3.14159);

d = orifice plate bore diameter calculated at flowing temperature T_f (feet);

g_c = dimensional conversion constant (lbm-ft/lbf-sec^2);

$\rho_{t,p}$ = density of the fluid at flowing conditions (Pf, Tf) (lbm/ft^3);

ΔP = orifice differential pressure (lbf/in^2).

Note: The square root symbol ($\sqrt{\ }$) and engineering exponent 0.5 ($Q^{0.5}$) are used interchangeably in this book.

In this equation q_m is the mass flow rate and is the value to be determined.

C_d is the coefficient of discharge that has been empirically determined; it was last revised in the 1992 edition of the ANSI/API Chapter 14-3/AGA-3 documents. It depends on the plate and meter run size.

E_v, the velocity approach factor, corrects for the change in the flow constriction shape with various beta ratio plates—as well as differential pressure and pressure effects—as the flow goes from a meter run to the orifice restriction.

Y, the expansion factor, corrects for the change in density from the pressure tap to the orifice bore in gas measurement. It can be calculated from either an upstream or downstream tap depending on which measures the static pressure. Its value is a function of the beta ratio, differential pressure, static pressure at the designated tap, and the isentropic exponent of the flowing gas. The factor is unity for liquid flows.

$\pi/4$ is used to convert the orifice diameter to an area; d is the orifice plate diameter as determined by proper micrometer readings.

g_c is a dimensional conversion constant.

$\rho_{t,p}$ is the density of the fluid at flowing temperature and flowing pressure.

ΔP is the differential pressure measured across the orifice.

All values must be in a consistent set of units such as those shown above.

Most standards reduce the basic mass flow equation (Equation 11.1) into one that allows the more convenient use of mixed units and reduces constants to a single number. However, all versions trace back to this general basic equation (Equation 11.1).

Each standard and each differential device has equations that look different (notations are not the same), but they follow a basic relationship. The fundamental orifice meter mass flow equation in the various standards (presented in Chapter 2 and repeated here for your convenience) is written in various ways:

API 14.3, Sec 3, Part 1

$$q_m = C_d E_v Y d^2 (\pi/4)\sqrt{2g_c \rho_{tp} \Delta P} \qquad (11.2)$$

AGA-3 1985

$$q_m = K Y d^2 (\pi/4)\sqrt{2g_c \rho_{tp} \Delta P} \qquad (11.3)$$

ISO 5167

$$q_m = CE\varepsilon d^2(\pi/4)\sqrt{2\Delta p\rho_1} \qquad (11.4)$$

ASME

$$q_m = (\pi/4)C\,\varepsilon d^2\sqrt{\frac{2\Delta p\rho_1}{1-\beta^4}} \qquad (11.5)$$

In these equations:
C = Cd
$E_v = E = [1/(1\text{-}\beta^4)]^{0.5}$
$Y = \varepsilon$
$2\Delta P = 2g_c\,\Delta p$
$K = C_d E_v = CE = c/[1/(1-\beta^4)]^{0.5}$
The equation may be simplified to use more easily measured variables (such as differential in inches of water rather than pounds per square inch), and some constants (such as $\pi/4$ and $2g_c$) may be reduced to numbers. If mixed units are used, corrections for these may also be included in the number. Details of this equation for each unit will be covered later in the appropriate section. Equation 11.1 can be converted to volume at base conditions by dividing the equation by the fluid density at base conditions:

$$Q_v = q_m/\rho_b \qquad (11.6)$$

Since flow measurement is often done with mixed units for convenience, the simplified equations contain a numerical constant for balancing the units. For example, in natural gas measurements, Equation 11.2 may be changed to:

$$Q_v = 218.573\,C_d E_v Y_1(d^2)(T_b/P_b)\sqrt{\frac{P_{fl}Z_{bair}h_w}{G_r Z_{fl}T_f}} \qquad (11.7)$$

where:
Q_v = volume flow rate (SCF/hr);
218.527 = numerical constant and unit conversion factor;
C_d = orifice plate coefficient of discharge;
d = orifice plate bore diameter calculated at flowing temperature (T_f) (inches);
G_r = real gas relative density (specific gravity);
h_w = orifice differential pressure in inches of water at 60°F;
E_v = velocity of approach factor;
Pb = base pressure (psia);
P_{fl} = flowing pressure (upstream tap) (psia);
T_b = base temperature (°R);
T_f = flowing temperature;

Y_1 = expansion factor (upstream tap);

Z_{bair} = compressibility at base conditions (P_b, T_b);

Z_{fl} = compressibility (upstream flowing conditions—P_{fl}, T_f).

Similar simplifications are also made in equations so applications are easier for standard units of the measured variables for liquid flows.

For these equations to be valid with minimum errors, the following factors are vital:

• The orifice plate and meter run must be kept clean and retain the original conditions specified by the standard.

• To ensure this, periodic inspection should be conducted to reaffirm conditions. Inspection frequency depends on problems of foreign material collection and possible damage. Inspections will confirm orifice diameter and thus the coefficient of discharge.

The expansion factor (Y_1) for gas provides a relatively small correction for high pressure (above 100 psia) measurement provided the differential, in inches of water, is not allowed to exceed the static pressure (psia). As this ratio increases (below 100 psia), the correction factor increases, as does the error in the factor.

Flowing density can be measured directly with a densitometer that is periodically calibrated. It also may be calculated from appropriate equations of state whose accuracy must be established from the data upon which the equation is based. In either case, density is usually the second most important variable to determine in the equation.

The most important variable in the equation—the one that primarily determines flow measurement accuracy—is the differential pressure. Therefore, a major effort should be made to have as high a differential as possible (considering flowing conditions), and the best available differential transducers should be specified. Anything that can be done should be done to improve this most critical measurement factor. However, the most important term in the equation is the orifice bore, d. This is the only term that is squared. Therefore, small errors or mis-measurement of its value will significantly impact calculated flow.

Maintenance

(The following supplements information previously presented in Chapter 9.)

Orifice meter maintenance consists of periodic inspection (as indicated above), cleaning primary elements, and scheduled testing and calibration against standards (if necessary) of the secondary elements. Maintenance frequency, if not set by agreement or contract, should simply be based on experience and performed as often as necessary to correct any calibration drift or error that may occur. Proper records for each station will determine this schedule.

Advantages of the orifice meter:

1. Well documented in standards;
2. Enjoys wide acceptance; personnel knowledgeable across the industry about requirements for use and maintenance;
3. Relatively low cost to purchase and install;
4. No moving parts in the flow stream; and
5. When built to standards' requirements, does not require calibration beyond confirming mechanical tolerances at the time of purchase and periodically in use.

Disadvantages of the orifice meter:

1. Low rangeability with a single device;
2. Relatively high pressure loss for a given flow rate, particularly at lower beta ratios;
3. More sensitive to flow disturbances at higher beta ratios than some meters; and
4. Flow pattern in the meter does not make meter self-cleaning (Figure 11-3).

Several special-shaped orifices for special applications are worthy of mention even though they are not in the industry's standards. The quadrant-edge (quarter circle) orifice is used at Reynolds numbers below which the square-edge concentric orifice coefficient becomes too non-linear to be useful. A conic-entrance orifice can be used for a similar range of Reynolds number; however, it can be applied for even lower than quadrant-edged Reynolds numbers. Both of these devices should be used with limited diameter ratios, and the required flow rates and Reynolds numbers must be evaluated carefully to ensure good measurement. (See Chapter 6, Reference 5.)

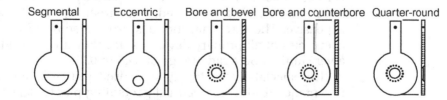

Figure 11-3 Many variations in orifice design allow for special measurement applications.

Eccentric, segmental, and annular orifices—with accuracies in the order of 2%—are special devices to take care of dirty fluids and two-phase flows. Since these are not the best devices for obtaining accurate measurement, they are used only where these special conditions exist. No detailed standards exist for these devices. For details on construction of these orifices, see Chapter 6, Reference 5.

Honed flow sections are orifice runs made in sizes of 0.25 through 1.50 inches. They are covered by the ASME, and data are available from manufacturers who have developed special manufacturing requirements and special coefficients to calculate corrected flow. These devices are used for low flows of gas, liquid, and steam with a higher tolerance than standard-sized meter tubes of 2.00 inches and larger that are covered in the industry standards.

Flow Nozzles

Another important flow element is the flow nozzle. Several configurations are available, the most important of which are the ASME long radius nozzle, high or low beta series, and the throat tap nozzle for gas and liquid. In Europe, another nozzle shape outlined is the ISA-1932 that is used more often than the ASME nozzle. Both have the same rangeability limitation as orifices—approximately 3 to 1 for a 100 inch recorder. (See ASME MFC-3M and ASME PTC-6 for details of construction.)

When flow rates change with time and thus require nozzles to be changed, this is more difficult than changing an orifice plate. The nozzle, however, is better able to sweep suspended particles by the restriction because its contour is more streamlined than the orifice. The ability to handle particulates is particularly good if a nozzle is installed in the vertical position with flow downward.

Nozzles are used mostly for high-velocity, non-viscous, erosive fluid flows. However, they have considerable acceptance in certain industries, such as electrical power generation. Standard nozzles are moderately priced, but throat tap nozzles are very expensive. The throat tap nozzles provide some of the highest accuracies of all primary devices, since they are allowed in only a very limited beta ratio range of 0.45 to 0.55.

The special throat tap type is used primarily for accurate acceptance testing of electrical generating plants. This is a very expensive nozzle because it must be flow calibrated prior to use, and its calibration must meet the standard coefficient value within ±0.25%. This does not allow much room for error in manufacture or calibration. Because of these problems, its use is primarily restricted to the power industry, where the acceptance testing can justify the nozzle cost (Figure 11-4).

The devices are very difficult to remove for inspection and cleaning, and their use in fluids when deposits may build up is not recommended. Installation requirements for nozzles are similar to those for orifices. Requirements are detailed in the ASME documents previously listed.

The nozzle has mistakenly been said to have low pressure drop. But for a given differential and pipe size, it is better stated that a lower beta ratio can be used. At times, a smaller tube size can be used with the nozzle than with the orifice. Permanent pressure drops will be approximately equal for a given set of flow conditions for either device if the differential range is set. Nozzle sizing must be based on good flow data that is fairly stable because of its restricted range. The expense of incorrect sizing should always be kept in mind.

ASME nozzle taps are located in the pipe wall one pipe diameter upstream and one-half pipe diameter downstream. The ISA-1932 nozzle uses corner taps. Nozzles are usually mounted between pipe flanges.

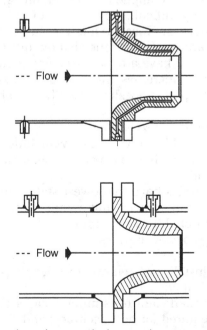

Figure 11-4 Flow nozzles—the top unit shows a throat tap, and the bottom unit represents an ASME "low beta" design.

Nozzles may be fitted with a diffuser cone to reduce the pressure loss by guiding flow back to the meter tube. It is possible to truncate this diffuser to approximately four times the nozzle throat diameter and have about as full recovery as with a cone extending to the wall.

The equation for the ASME (low or high beta) flow nozzle is in the same form as the orifice, with the ASME nozzle coefficient restricted to a Reynolds number of 10,000 or greater. The coefficient value is approximately 0.95 versus an orifice coefficient of approximately 0.60, so the nozzle handles over 50% more flow for a given size and differential reading. A special equation can be used for the coefficient at lower Reynolds numbers, but the change of coefficient in this range requires an iterative process of flow velocity and Reynolds number to maintain accuracy. Nozzles with other shapes may have different coefficient equations.

Operational accuracy of the nozzle is again directly related to the ability to measure at higher differentials because of the square root relationship with flow. It is best not to measure below 10 inches of differential, but higher differentials (400 to 800 inches of water, for example) can be used because of the mechanical strength of the nozzle versus the orifice.

The nozzle is best applied to clean fluids, since removal for cleaning is very difficult. Any change of nozzle throat finish has a direct effect on measurement accuracy. Where critical measurements are made, the installation must be capable of being shut down, or have a bypass to allow periodic inspection and cleaning. This increases the cost of a nozzle installation. For the most part, nozzles usually receive very little maintenance.

Advantages of flow nozzles:

1. Can be used at higher velocities with little damage to the nozzle surface, which allows using smaller sized meters for a given flow; and
2. Throat tap nozzles have the lowest stated uncertainty tolerances of all head devices, but require calibration to prove and have restricted beta ratio ranges.

Disadvantages of flow nozzles:

1. Expensive;
2. Difficult to install and remove for cleaning—and therefore are seldom cleaned;
3. Database for coefficient tests much smaller than for orifices; calibration required for best accuracy; and
4. Limited range of Reynolds numbers.

Venturi Meters

Low pressure drop devices are headed in the United States by the classical Venturi, which is used for liquid, gas, and steam at pipe Reynolds numbers of 100,000 and over. Venturi rangeability, like an orifice meter, is dependent on the differential readout and is 3 to 1 for a single 100 inch range recorder. Venturis are mainly used to measure liquids, clean or dirty.

The Dall tube is popular in the United Kingdom as a low-pressure device for these same measurements (Figure 11-5).

Where low pressure drops are required in non-viscous fluids, the pressure drop depends on the angle of the downstream cone and the beta ratio of the Venturi. Venturis have the disadvantage that their size makes inspection and changing cumbersome. It is recommended that an initial calibration be run for precision measurement, since they tend to be more difficult to manufacture precisely than a nozzle or an orifice. These problems can be secondary to the savings achieved in the costs of low operating pressure drops.

The classical (Herschel) Venturi is built to specifications of the ASME MFC-3M and ISO-5167 with an angled inlet and exit cone on a cylindrical throat section. It is designed to minimize pressure drop and to be somewhat self-cleaning, since there are no sharp corners in the flowing stream where materials may collect. Venturis have an overall length of approximately eight diameters; this makes them unwieldy to make, install, and remove for inspection. The actual design has several options

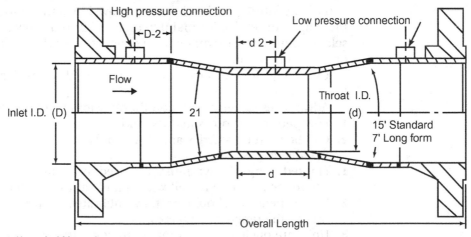

Figure 11-5 Herschel Venturi.

that must be chosen before the length and shape dimensions are set. One sometimes overlooked is inlet cone finish, which affects coefficient limits. Each Venturi is unique, and consequently the volumes to be measured must be accurately known, because once a Venturi is built, there is no flexibility for change of design flow rates in the unit.

Venturi Installation

Installation requirements for piping depend upon the upstream fitting (i.e., elbows, valves, reducers, and expanders) but are generally shorter than for a nozzle or orifice with the same beta ratio. No straightening vanes are required by the standards. However, experience shows that no swirl can be tolerated coming into a Venturi, so two elbows in different planes or pinched valves should not be installed in upstream piping, since they tend to create swirl.

In summary, Venturis are 3-to-1 flow range devices with a single 100 inch manometer recorder. Their throughput has the highest efficiency of any of the head devices with an approximate coefficient of 0.98. They should be designed to the actual conditions of flow with no design allowance for errors, thus the flow specifications must be correct.

The flow equation is the same as for the orifice previously covered except for the coefficient of discharge. A Venturi should perform correctly over time provided the surface of the inlet cone and the throat are not changed by corrosion, erosion, or deposits. Venturis can be cleaned if provision is made to remove the unit from service for maintenance. Other than this (which is seldom done), most maintenance consists of work with the secondary equipment.

Other low loss devices are the Universal Venturi and the LoLoss tube.

Advantages of Venturis (and other low loss head devices):
1. Low permanent pressure loss; and
2. Can be used on slurries and dirty fluids.
   Disadvantages
1. Limited rangeability; must be used only on installations where the flow rate is well known and varies less than 3 to 1.
2. Very expensive; should be flow calibrated to provide accuracy in the range of $\pm 1.00\%$; and
3. Units are big and weigh more than comparable head devices; this makes them difficult to install and inspect.

Other Head Meters

Pitot tubes (standard and averaging) can be covered in a single section since they are similar in their application limitations. They are used for measurement of liquids and gas. The meters do not "look" at the total flow profile, and their accuracies relate to how well the readings represent average velocity. When line velocity at a single point is measured, accuracy is generally no better than ±5% unless specific calibrations are run (Figure 11-6).

Over limited ranges, these devices are used for control purposes rather than custody transfer. Their major advantages are that they create practically no pressure drop and are inexpensive. On the other hand, velocities in normal pipeline designs are such that measured differentials in Pitots (which may be a maximum of 8 inches of water differential) limit application to fairly steady rates of flow (low rangeability). These very inexpensive devices may be installed so they can be removed for maintenance without stopping line flow.

Advantages of Pitot meters:
1. Low cost for installation and operation (essentially no pressure drop); and
2. Standard differential readout device for all sizes.

Disadvantages of Pitot meters:
1. Point velocity measured is assumed to represent full pipeline average velocity, which limits accuracy unless the flow profile is closely controlled; and
2. For normal pipeline velocities, the indicated differential is very low, which makes its accuracy poor and its rangeability very limited.

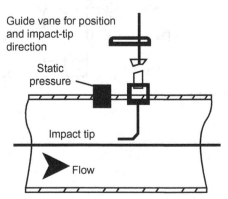

Figure 11-6 Pitot tube or impact meter.

Elbow Meters

As fluid flows around an elbow, centrifugal force makes the pressure on the outside wall higher than the pressure on the inside wall. This pressure difference is proportional to flow, and its coefficient can be estimated from a knowledge of the elbow dimensions. For more accurate measurement, an elbow (with at least ten diameters of straight pipe upstream—straightening vanes are recommended to stabilize swirling flow—and five diameters downstream) should be flow calibrated. If welds on an inlet elbow and pressure taps are carefully made, elbows will calibrate with a very stable calibration curve.

These units, however, are more often used for flow control (high repeatability) rather than for accurate flow measurement. Piping systems already have elbows present, and their use as a meter adds no pressure loss not already present. But normal pipeline velocities do not generate differentials (normal maximum about 9 inches), and this limits accuracy and severely limits rangeability (Figure 11-7).

Elbows are normally used for control with stable flow rates where repeatability is more important than absolute values of flow.

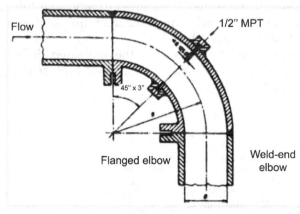

Figure 11-7 Elbow meter.

12

LINEAR AND SPECIAL METERS

A large number of meters used for measurement in the oil and gas industry are not head type meters. This chapter introduces these meters with a firm recommendation that they be closely studied for their particular application to your use.

Linear:

Non-Intrusive:

    Coriolis

    Magnetic (special purpose)

    Ultrasonic

Intrusive:

    Multiphase

    Positive displacement

    Turbine, gas and liquid

    Vortex shedding

Other and special purpose meters:

    Insertion

    Swirl

    Laser Doppler

    Nuclear magnetic resonance

    Sonic nozzles

    Thermal

    Tracer

Equations for non-head type volume meters are simpler, since they basically reduce the volume determined by the meter at flowing conditions to base conditions as covered in the section on basic laws.

$$q_f \, F_b = q_b \tag{12.1}$$

where:

q_f = volume at flowing conditions;

F_b = factor to reduce from flowing to base (corrects for effects of compressibility, pressure, and temperature);

q_b = volume at base conditions.

Fluid Flow Measurement. ISBN: 978-0-12-409524-3

If the meter measures mass, then there is no reduction to base required, and

$$q_{mf} = q_{mb} \tag{12.2}$$

where:

q_{mf} = mass at flowing conditions;
q_{mb} = mass at base conditions.

Non-Intrusive Meters

Coriolis Meters

These meters can be used on liquids and most gases. They directly measure weight (mass). If the desired measure is volume, then some correction for density at fluid base conditions must be made. Most models of the meter offer both mass rate and density *for liquids* from one device. Since these meters react to mass, they can be used (within limits) for some mixtures of liquids and gases. They do *not* measure gas density accurately. The manufacturer should be contacted for recommendation on a meter's limits with mixtures, since such an application presents special problems. Special construction materials are provided to minimize problems with hostile fluids (Figure 12-1).

Meter operation depends upon the Coriolis effect produced by the earth's rotation. (This effect can be demonstrated by dropping an object from a height. Rather than landing directly below, its landing point will be slightly displaced.) As a fluid passes through a tube electro-mechanically forced into vibration, a Coriolis force is generated, which alters the tube's vibrational mode. The magnitude of the time differential between no-flow and flowing conditions for the tube (or between two tubes in a dual-tube design) relates directly to the mass of the flowing fluid.

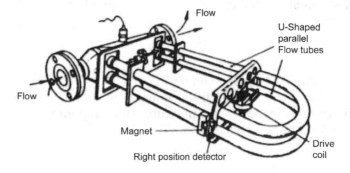

Figure 12-1 Modern industrial Coriolis meter.

Because the density of a gas is much lower than that of a liquid, the gas density measurement from a Coriolis meter is typically useful only as an indication and is not accurate enough for volumetric correction, which must be known or determined from some other source. A good combination for pipeline use is a Coriolis meter with a gas chromatograph (GC), since the GC provides both density information and heat (Btu) content—along with product-component identification and other information.

Coriolis meters are most popular in sizes of 1/16 inch through 6 inches. They are sometimes offered in larger sizes, but mechanical limitations begin to become onerous in these models. When a non-intrusive meter is desired in a size from 4 to 42 inches (or larger when required), another meter, such as the ultrasonic meter, should be used.

Coriolis meters are used for both custody transfer and control measurement. Their installed costs are similar to those of ultrasonic meters, and are relatively high compared to other choices. However, maintenance costs are minimal provided operation uses clean fluids. Any depositions or collections of materials in the meter body will be indicated as a density error. However, the high flow velocity tends to keep the meter swept clean. For this reason, single-tube Coriolis meters should generally be mounted vertically with flow downward. In most cases, dual-tube meters should be mounted "tube down" for liquids and "tube up" for gas measurement. Once again, the designer is warned that "clean" pipeline gas seldom, if ever, exists in the field.

Several standards exist for Coriolis meters measuring liquids, and a standard for custody transfer of natural gas with a Coriolis meter has been prepared but not yet released as this book goes to press.

The advantages of Coriolis meters are:
1. Can be used on liquids, slurries, gases, and two-phase liquids and gas flows (within set limits);
2. Units measure mass directly, which is an advantage when mass measurement is desired (see point 2 in "Disadvantages" below);
3. Can handle difficult fluids (highly varying densities or phase mixtures) where other meters cannot be used. (Check with manufacturer for suggestions and limitations in such applications.)
4. They provide high accuracy and repeatability on liquid flow and density; accuracy comparable to other meters generally used on liquid flow;

5. High turndown ratios (up to 100 to 1 or better);
6. Independent of swirl and flow profile; no flow conditioning required; and
7. Pressure ratings, low flow limits, and noise immunity greatly improved in recent years by some manufacturers.

The disadvantages of Coriolis meters are:

1. They are available only in sizes of 1/16 through 6 inches;
2. If volume measurement is desired, conversion via analysis or density measurement at base conditions is required; and
3. Special installation requirements are needed to isolate some meters from mechanical vibration.

Magnetic Meters

These meters are useful to measure *conductive* liquids or slurries. Because of the material conductivity they require for operation, they are not used in the petroleum industry for measuring hydrocarbons (Figure 12-2).

They are very useful, however, for measuring such things as water slurries, which most other meters cannot measure. They are made in sizes from fractions of an inch through almost 100 inches. Since they operate on velocity, the equations needed to convert from flowing to base conditions are the same as for turbine meters.

Density and viscosity do not directly affect meter operation. The meter operates bidirectionally, provided upstream lengths are used on both sides of the meter to control the velocity pattern. Since the meter is full line size, there is no pressure drop caused by the meter other than normal pipe loss. The meters are fairly expensive and have a high operating cost because of their high power requirements. The larger sizes are quite heavy and require special considerations for installation and removal.

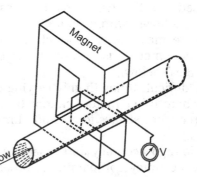

Figure 12-2 Elements of an electromagnetic flow meter.

The advantages of magnetic meters are:
1. Performance not affected by changes in viscosities and densities;
2. Full-bore opening means no head loss;
3. Meters will operate bidirectionally with required upstream lengths installed on both sides of meters; and
4. Available with insert liners that allow use on some corrosive and erosive fluids.
  The disadvantages of magnetic meters are:
1. Installation and operating costs relatively high because of the size, weight, and cost of electrical power;
2. Fluids must have at least the minimum conductivity specified by manufacturer of the specific meter; and
3. Used for liquids or slurries but *not* gases.

Ultrasonic Meters

The ultrasonic meter category contains a number of different designs for measuring an average velocity in a flowing system. They are all based on an ultrasonic signal being changed by or reflected from the flowing stream. Meter accuracy relates to the ability of the system to represent the average velocity over the whole stream passing through the meter body's hydraulic area. This ability affects installation requirements and the accuracy of results obtained.

Dopplers

The two main types of ultrasonic meters are the Doppler frequency shift and the transit time change. The Doppler meter is used on liquids and gases with some types of entrained particles that are traveling at the same speed as the main body of flow. The ultrasonic signal is reflected from these traveling particles across stream, and the shift in frequency is related to the average velocity of these particles over time. Meters are made in several types; one type requires installation of transducers into the flowing stream, the other is a strap-on model that can be installed without shutting down the flow stream (Figures 12-3, 12-4).

Transit Time Ultrasonic Meters

A transit time unit is installed directly into the flowing stream and can be made with single or multiple transducers for establishing average velocity. These units can be used on liquids or gases, although the great majority of pipeline applications

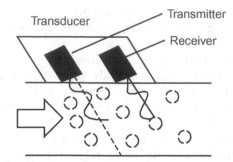

Media with scatters (dirt, air, etc.), and ultrasonic wave reflect

Figure 12-3 Doppler reflection meter.

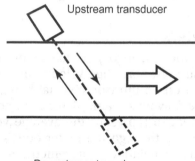

Each transducer alternately acts
as a transmitter and receiver.

Figure 12-4 Transit time ultrasonic meter.

are for gases. The multiple transducer units can handle velocity profile distortions so that installation requirements are reduced. But meter complexity (i.e., cost) goes up because of the multiple transducer units and the more complex electronics required to compute average velocity and flow. Some manufacturers offer spool piece single and multiple path meters plus insertion ("hot tap") types for surface or underground installation.

The multipath meter uses transducers set at an angle to the flow axis. In one company's four-path design, each transducer in a pair functions alternately as transmitter and receiver over the same path length. When equations for transit times "upstream" and "downstream" are used to determine mean transit time, the speed of sound in the medium "drops out." Consequently, gas velocity through the meter can be determined from only transit times and the physical dimensions of the spool piece (Figure 12-5).

Equations

Volume flow rate equals weighted, calculated mean velocity times meter bore cross-sectional area. To convert from flowing volume at base conditions, corrections must be made for pressure and temperature as for a turbine meter (Figure 12-6). Here

Figure 12-5 Transducers in this multipath ultrasonic meter are mounted so that the sound travel path crosses the pipe body at a set angle. Four pairs of transducers are used for high accuracy with many flow profiles.

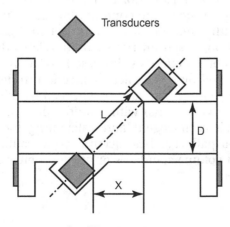

Cos "Theta" = X/L

Figure 12-6 Schematic of one configuration of a "direct-path" ultrasonic meter showing critical angles and dimensions.

are the equations involved for one configuration of a transit time meter:

$$t_{ud} = \frac{L}{C + V_m \cos \Theta} \tag{12.3}$$

$$t_{du} = \frac{L}{C - V_m \cos \Theta} \tag{12.4}$$

$$V_m = \frac{L}{2 \cos \Theta} \left(\frac{t_{du} - t_{ud}}{t_{du} {}^* t_{ud}} \right) \tag{12.5}$$

Actual volume flow rate (Q) is

$$Q = V_m \left(\frac{\pi D^2}{4} \right) = K \left(\frac{t_{du} - t_{ud}}{t_{du} {}^* t_{ud}} \right) \tag{12.6}$$

where:

t_{ud} = transit time from transducer U to D;
t_{du} = transit time from transducer D to U;
L = distance between transducers U and D (path length);
C = velocity of sound in the gas;
V_m = mean velocity of the flowing gas;
Θ = angle between acoustic path and meter axis;
D = diameter of the meter bore;
K = constant for a specific meter application.

Performance

An ultrasonic meter's performance depends on its ability to find the average velocity, the condition of the meter's open area (no change with rate), and the abilities of the readout system. Meter calibration, based on transit time, relates directly to the mechanics of construction, as discussed above. It can be calculated and checked by mechanical inspection to determine geometrical dimensions. It can also be checked by filling the meter (during no-flow conditions) with a fluid whose speed of sound is known (such as nitrogen) and calculating the transit time over the signal path. For best accuracy, the calibration should be run against volume standards or master meters as covered in AGA Report #9.

Maintenance

There are no moving parts that require lubrication, and maintenance is basically performed only on the readout system for meters installed to measure clean fluids. For dirtier fluids,

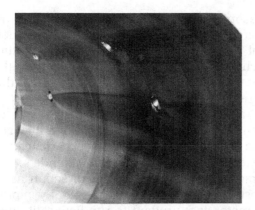

Figure 12-6a An apparently clean gas ultrasonic meter showing upstream transducer wake impingement on the face of the downstream tranducer.

cleaning must be done if the meters' flow areas or transducers are affected. The meter causes no pressure drop across it other than the normal drop in an equivalent length of pipe. The meter has a fairly large turndown ratio with accuracies for multipath designs equal to the best of other types of meters (Figure 12-6a).

Multipath meters with greatly extended low-flow accuracy (accurate measurement down to flow velocities of 1 ft/sec rather than the typical limit of 5 ft/sec) have been developed. Utilization of this capability was greatly enhanced in 2000 when a major US testing facility automated its low-flow calibration facilities. The extended rangeability allows one meter to suffice instead of having to install a second smaller meter for low-flow measurement.

However, the low-flow (1 ft/sec) accuracy that is achievable in the flow laboratory is questionable in the field environment. Meters operating at very low fluid velocities are subject to adverse ambient temperature fluctuations resulting in possible temperature stratification within the meter and bias temperatures used for standard volume correction, 0.2% per °F bias. Covering the metering installation can minimize this influence by removing the radiant heat transfer element but will not protect the metering from the conductive and convective heat transfer environment of an infinite ambient source; see Chapter 16 for additional information.

Single path designs are more sensitive to flow pattern irregularities and provide less accuracy without flow profile stabilization. Data on the accuracies of these meters are available from standards, independent laboratories, users, and manufacturers.

The ultrasonic principle is applicable to all pipeline sizes; manufacturers' literature lists available meter sizes. Bidirectional flows can be measured with no additional electronics, mechanics, or piping configurations other than treating both sides as "upstream." The change of timing of the upstream and downstream signal reflects the flow reversal ("positive" or "negative"), and the electronics calculate flow accordingly. There is a small amount of dead space when flow goes through zero and the minimum velocities are experienced. This "dead band" is obviously greatly reduced with extended low flow capability. Accuracy is the same in either direction as long as the minimal upstream requirements are met by piping on both sides of the meter. It is prudent to run flow calibrations in both directions for highest accuracies. These requirements change depending on the type of ultrasonic meter used. These meters are available in wide temperature and pressure ranges.

Advantages of ultrasonic meters:

1. No added pressure drop, since meters are the same diameter as adjacent piping;
2. If a high frequency pulse rate of output is used, it can minimize errors from effects of pulsation and fluctuating flow;
3. Installation can be simple and relatively inexpensive;
4. High rangeability (the highest with extended low-flow capability);
5. No moving parts in contact with flowing fluid; and
6. Simple mechanical calibration easily checked.

Disadvantages of ultrasonic meters:

1. Power required for operation;
2. For a single path or reflection unit, the flow profile must be fully developed for an average velocity to be determined (note: multiple path unit's average disturbed flow patterns including small swirls to minimize flow profile problems); and
3. High initial cost.

The throughput test required by the AGA-9 standard should be conducted for highest accuracy.

Intrusive Linear Meters

Multiphase Meters

A single meter to measure the rate of three-phase flow—oil, gas, and water—has been sought in the oil and gas industry for over 20 years. Use offshore is especially appealing, since such a meter would eliminate the cost of platforms with the space and weight of conventional systems with test separators, associated piping, and individual meters for each phase. In cases in which economics control, and it is simply not practicable to build the platform for conventional equipment, a multiphase meter might represent the only approach available to allow a small reservoir to be produced.

Multiphase metering is not a simple task. Mother Nature mixes and separates the gas, oil, and water in a multitude of variations depending on pressure, temperature, specific velocities, gas and oil compositions, specific gravities/densities, viscosities, thermal characteristics, water salinities, volumetric fractions GOR/GLR, etc. Down-hole chemical injections also play a part in the ways that the various fluids affect flow regimes and how the fluids are measured. All the variables influence what type of technology is best suited for a particular production scenario.

It is important to realize that there may be two accuracy "levels" associated with multiphase metering: (1) "good enough" for reservoir management, and (2) as accurate as conventional systems (or better) for custody transfer measurement. In each case it is important to compare the cost of multiphase measurement with the cost of conventional systems *for equivalent measurement accuracy.* Most potential users agree that "as good as or better" would be the goal for custody transfer measurement and meter acceptance in most offshore applications.

There are many approaches and technologies that have been tested as part of a multiphase flow measurement system. They include:

Volume meters:
   Venturi;
   Gravity meter;
   Coriolis; and
   Positive displacement.
Density meters:
   Microwave;
   Gamma ray;
   Dielectric permantivity; and
   Nuclear.
Homogenizers:
   Static mixers; and
   Cross correlations (from tests);
Partial separation:
   With high gas contents.

Meters currently being offered commercially may achieve, or be near to achieving, accuracies good enough for reservoir management. Custody transfer accuracy will require further development of the multiphase meter, although accuracy that matches existing conventional "meter-after-separation" is possible in some situations.

For example, one leading manufacturer is offering a gamma radiation system for measuring gas volume fractions (GVF) and water, combined with an integral Venturi for total (bulk) flow. The company currently claims the following accuracies:

Water cut error: ±2% absolute for 0 to 100% water;

GVF error: ±2% absolute for 0 to 30%, ±5% for 30 to 50%, and ±10% for higher percentages;

Liquid flow rate ±5% of reading; and

Gas flow rate: ±8% of reading.

Partial separation is used for GVF greater than about 25 to 30% and the separated gas is measured separately.

This particular meter has a small low-power electronics package for operation, and it provides real-time salinity measurement that is a variable. Salinity measurement not only allows fraction and flow compensation along with reducing the need for calibration, but can also be valuable to reservoir engineers concerned with water-injection or water-fractured wells. Salinity also affects the pressure and temperature conditions at which line-plugging hydrates can form.

Multiphase meters are currently being carefully evaluated by both land and offshore operators, with the latter having considerably better economic incentives. By the time this book is published, several trial installations may provide valuable guidance for further technology development.

Conclusions

The day of an operational multiphase meter is here for selected uses. Units are and have been installed in sites around the world. Acquiring a multiphase meter for field operation is like buying a pair of shoes. No one particular piece of equipment will fit all installation or operational criteria. You must know and verify the equipment's operational characteristics to be able to fit it into a particular field installation. All multiphase meters are dependent on high speed computers and are very software intensive. Manufacturers state accuracies and/or uncertainties in differing ways. Care must be taken in determining just what the capabilities of the instrument are and what degrees of accuracy can be expected. With further developments, the technology and performances will be better defined.

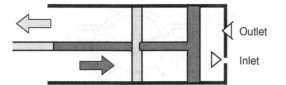

Figure 12-7 Cylinder and piston show principle of displacement metering.

Positive Displacement Meters

Positive displacement (PD) meters are used for the measurement of liquids and gases, primarily liquids for pipeline uses. Two of the most common meters at residences are the water and gas meter, both of which are usually PD meters (Figure 12-7).

The basic PD design is a "bucket" that is alternately filled and emptied. To keep operation from being strictly batch operation, most PDs have multiple "buckets" that are geared and valved together so that while some buckets are filling, others are emptying. With proper timing and valving, there is an uninterrupted flow through the meter. The driving force for this action comes from the flowing stream as a pressure drop.

Readout is typically achieved electronically, although direct gearing to a Veeder-Root type counter allows a PD meter to be used at sites where no external power is available. When stopped, most PDs create a large blockage factor in the line. If there is a danger that the meter may be jammed, some means of bypass or relief must be designed, otherwise flow may be reduced to a very small percent of normal. In some cases a severe flow reduction may not be a critical concern, unless the pressure drop builds to a point that major damage is done to the meter. This problem is somewhat unique to PDs and must be examined before metering system designs are set.

Rangeability

One characteristic of PDs that makes them attractive is a rangeability that is unmatched by most other meters. They are able to measure very small flows in their stated ranges with relatively high accuracy. Since the meter's flow path is normally not straight through, requirements of upstream and downstream piping are minimal and are normally based on piping space concerns rather than flow pattern considerations. Periodic maintenance and calibration must be allowed for in the design provisions to interrupt or bypass flow (Figure 12-8).

Positioned-vane rotary

BiRoto meter (double case)

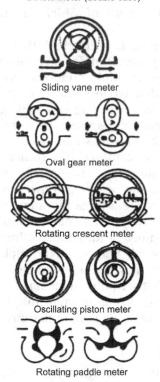

Sliding vane meter

Oval gear meter

Rotating crescent meter

Oscillating piston meter

Rotating paddle meter

Figure 12-8 Major types of positive displacement (PD) meters for liquid measurement.

Design

Design of PD meters varies from device to device, and serves various portions of industry needs depending upon operational characteristics. For example, common to the natural gas industry are four-chamber/two-diaphragm meters, three-chamber meters with an oscillating valve, rotary meters of the roots type, meters with four rotating vanes and rotoseal meters. Liquid PD types include reciprocating piston, ring piston, rotating discs sliding vane, rotating vane, oval gear, nutating disc, metering pumps, and lobed impeller.

Standards and manufacturers' literature should be consulted for operational details, sizing, range and accuracy limits, and recommended applications.

Because of the mechanical size of PD meters, limits on pressure, cost and weights become increasingly important as the meters get larger than about 10 inches. They are made in larger sizes, but the applications of these models are generally limited to special flow measuring situations.

Performance

PD meters perform well for long periods of time provided the fluids are clean, non-erosive, non-corrosive, non-depositing, and proper maintenance is routinely performed. Most major manufacturers also offer special designs for "hostile" fluids. In less critical measuring systems, such as domestic water and gas, meters are run for many years with testing only on the basis of statistical failure-rate study or upon customer complaint.

On the other hand, large liquid PD meters used in the petroleum industry may be tested on a weekly basis with a prover system permanently installed as part of the metering station.

There should be no overranging; if necessary, a protective flow-limiting device (with automatic bypass valve) should be installed to prevent the overranges, which mechanically damage the meter.

Equations

The equations for PD meters are the same as for turbine meters. Readout of the PD, which is at line volume conditions, is multiplied by a meter factor arrived at from proving and correcting factors to reduce flowing conditions of pressure, temperature, and compressibility at higher pressures to the base or contract conditions. (See Equations 12-7 through 12-9.)

Maintenance

Maintenance of small PD meters (below 4 inches) usually involves replacement, with repairs or rebuilding done at a central meter shop rather than in the field. With larger meters (6 inches and above), maintenance consists of part(s) replacement in the field and removal to a meter shop only if field repairs are unsuccessful. Maintenance decisions are based on the economics of maintenance costs versus the cost of inaccurate measurement.

Advantages of PD meters:

1. Insensitive to upstream and downstream piping effects so that no, or minimum, lengths are required;
2. Operating principle straightforward, easy to understand;
3. Rangeability among highest of liquid and gas meters available without loss of accuracy;
4. Even though valving and clearances require close tolerances, commercially available units are rugged and provide long and reliable service on clean fluids or with line filters; and
5. Simple to complex readout systems available for a simple flow equation.

Disadvantages of PD meters:

1. Because of clearances required, pressure, temperature, and viscosity ranges are limited and special care may be required for installation (meter specifications in this regard vary between manufacturers and should be examined carefully);
2. For larger sizes (above 10 inches), meters are large, heavy, and relatively expensive;
3. Head loss can be high, particularly if the meter jams; protection from flow shutdown and pressure overrange may be required;
4. Filtration or strainers may be required for fluids containing foreign particles to minimize meter wear; and
5. Maintenance costs are high on some larger meters; unit replacement is typical for smaller meters because of their complexity and field-repair cost.

Turbine Meters

Turbine meters are used successfully and widely in both liquid and gas measurement. They are made differently for gas and liquid measurement because of the difference in driving forces of the fluids and internal bearing frictions. However, the basic operation is the same for gas or liquid service.

The turbine meter is a velocity measuring device. Flow passes through a free-turning rotor mounted coaxially on the meter body centerline of the body. Since velocity is the parameter measured, the upstream downstream piping must have defined lengths to eliminate nonstandard velocity profiles and swirl (Figures 12-9, 12-10).

Fluid imparts an angular velocity to the angled rotor so that the rotation is proportional to the flow rate. Blade shape and angle, bearing style, and other construction details vary from manufacturer to manufacturer. With accurate measurement of rotor speed from mechanical gearing or magnetic pickup, and by knowing the hydraulic area that the flow is passing through,

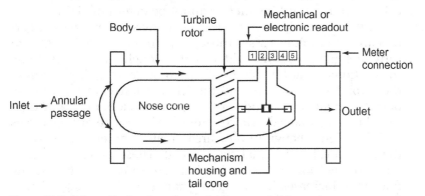

Figure 12-9 Schematic drawing of an axial flow gas turbine meter.

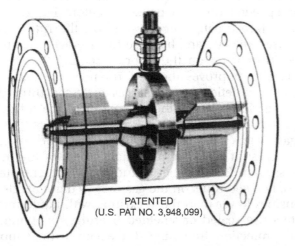

Figure 12-10 Typical liquid turbine meter. Its internal design is such that upstream and downstream forces on the rotor balance for a "floating" rotor.

the volume at line conditions can be determined. Since no rotor is without friction, the aerodynamic friction across the rotor and the friction of the bearing system cause a non-linear calibration until the retarding forces are a small percentage of the driving forces. At this point, the calibration curve becomes linear (rotor speed increases directly with flow velocity).

The actual flow area is not the calculated open area, but something less because turbulence blocks some of it. Because of the friction, turbulence, and necessary manufacturing tolerances, each meter must be calibrated to determine its proof curve. The necessity for this calibration is different than for an orifice that can be manufactured to a tolerance that allows its calibration to be predicted from its mechanical shape; precise calibration for a turbine meter cannot be determined this way.

As flow through the turbine first increases from zero, a certain amount of fluid passes through the rotor before it begins to turn. At some point, the fluid imparts enough force to overcome the frictional retarding forces of the rotor bearing. At this point, the rotor begins to turn, and the friction forces in the bearing become small. The aerodynamic forces predominate and control the rotor's speed. The existence of these retarding forces and the slight change in flow area create a difference between the theoretical and actual rotor speed. These differences must be accounted for with a calibration run on each meter. As the flow rate increases, these aerodynamic and bearing friction forces become minimal, and the proof curve becomes linear, reflecting only an increase in velocity.

There is another kinetic effect to consider. Fluid entering the meter is speeded up by a deflector before it passes through the rotor. More driving force results on the rotor because of the increased velocity and because the average velocity is being imparted further out on the rotor increasing the lever arm of the force. This improves the performance curve at lower flow rates. The flow deflector also serves to lessen thrust loads on the rotor bearing by shielding the center of the rotor from the flowing stream.

For gas meters, the deflector is larger (and the annular opening smaller), so it blocks approximately 66% of the meter area versus less than 20% for liquid meters. This generates higher velocities, hence torque, on the gas meters that operate in fluids with densities lower than those existing with liquid meters. The gas meters use sealed or shielded bearings for minimum friction and protection from line dirt accumulating; most liquid turbines use sleeve bearings of tungsten carbide for rugged wear characteristics.

Rangeability

The rangeability of a gas turbine meter varies with pressure—approximately 10 to 1 on a gas at atmospheric pressure to over 100 to 1 on a gas at pressures over 1,000 psia. On the other hand, liquid meters maintain a constant range of approximately 10 to 1 but have some overriding concerns for changes in viscosity, density, and meter size. As the viscosity rises above that of water, the meter range can be diminished down to as little as 3 to 1. Likewise, as densities drop to 60% of the density of water and lower, the range begins to decrease until it may reach 3 to 1. Smaller sizes (below 6 inches) tend to have lower ranges of linear proof curves. Each specific manufacturer should be consulted about the degradation of a meter range with viscosity, density, and size.

Meter capacity is determined by allowable rotor speed (bearing speed limit), pressure drop, and fluid velocity (blade angle). All manufacturers choose different design parameters, so their specific meters handle volumes which may be similar but not equal to that of another manufacturer (Figure 12-11).

A. Normal installations

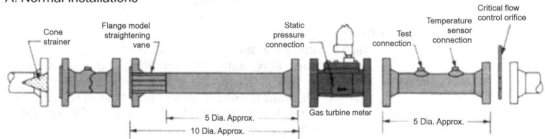

B. For maximum accuracy in uncertain installations

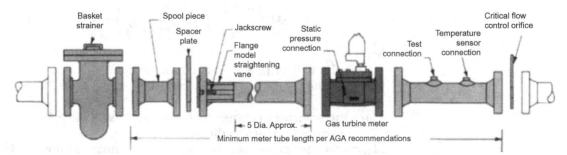

Figure 12-11 Installation of two- and three-section gas turbine meter tubes according to the AGA-7 requirements.

Installation

Installation of a gas turbine must be done according to AGA-7 or ISO-9951 (draft). Liquid turbine meters must be installed according to Chapter 5 of the API *Manual of Petroleum Measurement Standards* for custody transfer metering. Other standards are more lenient in required lengths. Piping should be designed to allow for testing and removal of the meter for repairs as necessary, since neither liquid nor gas meters can be removed from an operating line without stopping flow and depressurizing. If the delivery requires flow continuity, then a bypass must be installed. (Note: The laws of some countries require metering continuity on any custody transfer meter.) If this is the case, then the bypass must include a meter. Sizing tables supplied by the meter manufacturer should be used in designing a meter station. If flow rates fluctuate, the range of the flows should be maintained within the turbine meter limits, particularly for low flow rates where meters have larger errors if the lower limits are exceeded.

Equations

The general turbine meter equation is:

$$q_b = q_f M_f \left(\frac{\rho_{t,p}}{\rho_b} \right) \tag{12.7}$$

where:

q_b = volume flow rate at base conditions;
q_f = volume flow rate at operating conditions (meter reading);
M_f = meter factor to correct meter output based on calibration;
$\rho_{t,p}$ = flowing density at operating pressure and temperature;
ρ_b = base density at reference conditions

Equations for gas and the liquid meters are different. The gas turbine meter equation is as follows:

$$q_b = q_f M_f \left(\frac{P_f T_b Z_b}{P_b T_f Z_f} \right) \text{ and } \left(\frac{P_f \, T_b \, Z_b}{P_b \, T_f \, Z_f} \right) = \frac{\rho_{t,p}}{\rho_b} \tag{12.8}$$

where:

q_b = volume flow rate at base conditions;
q_f = volume flow rate at operating conditions (meter reading);
M_f = meter factor to correct meter output based on calibration;
P_f = pressure flowing conditions;
P_b = base pressure set by agreement near atmospheric pressure;

T_b = base temperature set by agreement at 60°F;
T_f = temperature at flowing conditions;
Z_b = compressibility at base pressure and temperature;
Z_f = compressibility at flowing pressure and temperature.
The liquid turbine meter equation is as follows:

$$q_b = q_f\, M_f\, F_t\, F_p \text{ and } F_t F_p = \frac{\rho_{t,p}}{\rho_b} \qquad (12.9)$$

where:

q_b = flow rate at base conditions;
q_f = flow rate at operating conditions (meter reading);
M_f = meter factor to correct output based on calibration;
F_t = factor to correct fluid from flowing temperature to base temperature;
F_p = factor to correct fluid from flowing pressure to base pressure.

A turbine meter operates over its specified range with equal accuracy. Overranging by pressure drop can damage the blades, or high velocity can damage the bearings. This is a particular problem while putting meters in and taking meters out of service; at these times, flow rates must be changed slowly. Meters used with liquids that vaporize as pressure is removed from them require special filling techniques so that the meters are not damaged. This can be done by slowly filling the system with a gas while monitoring the rotor speed. Filling is continued until the pressure of the liquid is reached. The gas can be bled off slowly while the liquid is allowed to displace it without pressure drop and vaporization. When the liquid has completely filled the system, liquid flow may be started.

Maintenance and Calibration

Maintenance for properly operated turbines consists of periodic cleaning and physical inspection. Calibrations may be required to reconfirm proof curves on custody transfer meters. This may be done by calibration against standardized master meters or direct calibration against standards (i.e., critical flow nozzles for gas; pipe provers for liquids and gas).

Advantages of turbine meters:

1. Good accuracy over full linear range of meter (accuracy is percent of flow rate, not percent of full scale);
2. Electronic output available directly at high resolution rate which makes proving possible in a short time period with smaller prover time or volumes;

3. Meter cost is medium, but total meter station is low-to-medium cost because of high flow rate for given line size;

4. Has pressure and temperature limits, but can handle normal flow conditions very well; and

5. Excellent rangeability on gas meters at high pressure.
   Disadvantages of turbine meters:

1. Require throughput proving to establish most accurate use;

2. Viscosity affects liquid meters that may require separate proof curves for different viscosities;

3. Rangeability at low pressures about the same as other gas meters; and

4. Require upstream flow pattern to be non-swirling, which necessitates straightening vanes.

Vortex Shedding Meters

The vortex shedding meter has come into prominence and usage in the last 20 years for both gas and liquid measurement. It has received acceptance in the industrial flow measurement area and, to a limited degree, in the custody transfer measurement area (Figure 12-12).

Although based on the same basic principle, various vortex shedding meters are standardized by performance, not in terms of mechanical construction of primary and secondary elements of the meter.

Operation

The vortex shedding meter operates on the Von Karman effect of flow across a bluff body. This principle states that flow will alternately shed vortices from one side and then the other of a bluff body, and the frequency of shedding is proportional to velocity across the body. When this velocity is combined with the hydraulic area of flow in a stream, the rate of flow can be established. Action is similar to the movement of a flag

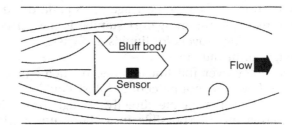

Figure 12-12 Major components in a vortex shedding meter.

downstream of a flag pole. The rippling of the flag is due to the vortices as they are shed alternately on each side of the flagpole. Vortices may be counted in many ways, since the vortex represents a pressure and temperature change, and either of these may be sensed. Or, a secondary effect of the small movement at the bluff body can be used.

In any case, vortices are shed irregularly at low flows. When these stabilize, the meter's lower flow rate is defined. Manufacturers have made continual developments in the readout systems and in determining the bluff body shape to give a strong, stable shedding pattern. Because of individual differences in the bluff body and the readout, each design is unique, and meter calibration should be obtained from the manufacturer.

Since the meter reacts to velocity, it follows that a proper flow pattern must be presented to the bluff body. This is accomplished by using straightening vanes, flow profile generators, and/or straight upstream piping to eliminate swirl distorted patterns. Installation requirements are similar to other velocity sensitive meters.

Sizing

The sizing of these meters with normal pipeline velocities makes throughput per line size higher than many other meters. Manufacturers' sizing recommendations should be followed.

Equations

The equations for vortex flow meters are the same as those for turbine meters (Equations 12-7 through 12-9), since the meter produces a pulse output proportional to the flow rate at line conditions, and this output must be corrected from line conditions to base conditions. Depending on the individual meter, a calibration factor K is determined, which relates produced pulses to the line volume passed versus Reynolds number. These factors are supplied by the manufacturer based on calibrations covering a range of Reynolds numbers (liquid and gas) similar to the operating Reynolds numbers. The K curves are quite linear for flows above the low-end limit. Viscous liquid should be checked to make sure its Reynolds number at flowing conditions will exceed the low flow limit, usually in the 10,000 range.

Maintenance

As long as the bluff body and the meter body opening are kept clean, the meter should retain its original calibration. Any erosion, corrosion, or deposits that change the shape of the bluff body will cause a change in hydraulic area and will shift calibration. Periodic inspection is recommended to ensure that initial conditions are being maintained in the primary element.

The flow-variable correction instruments of the secondary system must be calibrated to insure that the transducers have not changed calibration. If recalibration of the primary element is required, then some type of throughput test is run against a standard.

Advantages of vortex shedding meters:

1. Relatively wide rangeability with linear output;
2. On clean fluids (liquids and gases), the meters have long-term stable proofs;
3. Frequency output can be read directly into electronic read-out systems;
4. Installation costs moderate; installation simple;
5. When minimum or higher Reynolds numbers pertain, effects of viscosity, pressure, and temperature are minimal; and
6. No moving parts in contact with the flowing stream.

Disadvantages of vortex shedding meters:

1. Flow into a meter must be swirl-free; this requires straightening vane and/or long, straight piping;
2. Output may have "jitter" (frequency instability) and/or fade in certain areas of operation which affect readout requirements;
3. Not available in sizes above 8 inches;
4. Pulse train is irregular, proving requires a long test time to obtain a representative average pulse rate;
5. Pulse resolution the same for all meter sizes; this means a low pulse rate with larger meters yields low volume resolution; and
6. Subject to range limitations at lower Reynolds numbers.

Other and Special Purpose Meters

In addition to the more common meters covered above, other meters are available that are worthy of mention and a short description. When an application is contemplated, specific information should be obtained from suppliers of the meter.

Insertion meters are available in three types based on previously covered meters: turbine, vortex, and magnetic. An insertion meter is used to sample a velocity representative of the average velocity of a full stream. This limits their absolute

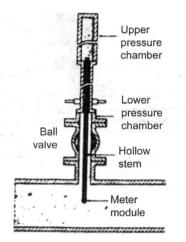

Figure 12-13 Insertion-type turbine meter.

accuracy to the validity with which the velocity sample point is located. However, their repeatability may be sufficient for some uses as a control device. Because of the small amount of flow blockage, insertion meters cause zero to small pressure drops compared to drops of equivalent full-bore meters (Figure 12-13). (Refer to previous material in this chapter for application suggestions and limits for the three meter types upon which insertion meters are based.)

Swirl meters have a fixed-geometry helix blade, which imparts a swirl upstream of a Venturi-shaped throat that increases the stream velocity. Deceleration in an expanding cone then follows. This action generates a precessing vortex (swirl) whose frequency is a function of flow velocity through the meter. A sensor then picks up the temperature change in the swirl or variation in pressure caused by the swirl. The meter is normally used on gas flows because of its high head loss on liquids. The complexity of mechanical construction involved requires calibration for each meter for best accuracy (Figure 12-14).

Special Application Meters

Certain flow meters have specific applications but are not considered for general use. The exclusion may be due to cost, or because the meter is newly developed or limited by its design/operation. This does not mean that at some future time there may not be further developments to expand a meter's use.

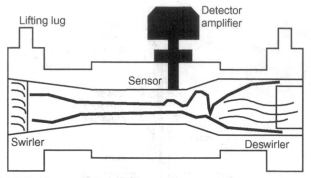

Figure 12-14 Swirl meter schematic.

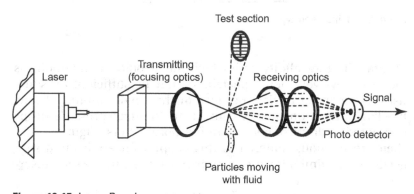

Figure 12-15 Laser Doppler anemometry.

Such meter types include thermal, tracer, laser Doppler, nuclear magnetic resonance (NMR), and sonic nozzles.

Laser Doppler meters are similar in concept to Doppler ultrasonic meters but are more expensive and difficult to set up. For these reasons, they are more often used in research facilities for flow measurement. They require transparent pipe and flowing fluids that allow light to penetrate the flow stream (Figure 12-15).

Two light beams are focused on a particular area where flow velocity is to be measured. Any light-sensitive particles that pass this point scatter the light, which is then measured by a photo detector. The particle velocity causes a Doppler shift that produces a signal in the detector proportional to flow velocity at that point. Beams are moved across the flow to enough points to establish an average flow velocity.

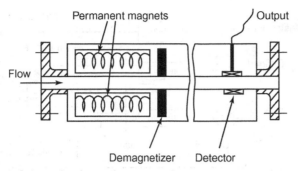

Figure 12-16 Schematic of a nuclear magnetic resonance meter.

These devices have definite use in studying flow profiles and patterns but have little use in converting a flow pattern to flow volume.

NMR meters mark the nuclei of hydrogen or fluoride in a flowing fluid. The fluid then enters a detector section when the magnetized nuclei relax between two detectors. This measurement produces a frequency proportional to fluid velocity within the detector section whose length and volume are known (Figure 12-16).

These devices, which are still in developmental infancy, are very expensive. However, their unique characteristics allow them to cover a wide range of difficult-to-measure flows (i.e., slurries, non-Newtonian fluids, and emulsions). Industrial applications are limited; most uses are in development and research work.

Sonic nozzles are specially shaped nozzles used for calibrating gas flow devices. A sonic nozzle is used mostly as a test device, because its high permanent pressure drop (10 to 15% of inlet pressure) is too costly to absorb in operations on a continual basis. It is quite accurate, provided that the thermodynamic properties of the flowing gas are known accurately. The most common use is as a calibration device for natural gas meters, such as PD or turbine meters used at over 35 psia pressure. Details on meter construction, calculation procedures, and use are available from meter manufacturers. The nozzles measure only one flow rate for a given static pressure and therefore are not used for normal flow measurement.

Thermal meters have been used for some research applications. However, developments are now beginning to make them attractive for some commercial (non-custody) flow jobs. These meters follow two basic operational principles:

1. A body is heated by known heat input, and the body is cooled by the flowing stream; this temperature change is proportional to mass flow rate.

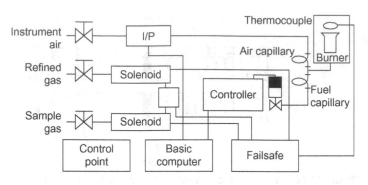

Figure 12-17 Thomas thermal flow meter.

2. A heat source adds heat to the stream, and downstream sensors measure the temperature rise which is proportional to mass flow rate (Figure 12-17).

The hot-wire anemometer is used to define flow velocity in a clean gas stream flow as the fluid cools the element. The element is operated at constant current, and its resistance is kept constant so voltage variations measured relate to velocity. In either case, the sensitivity of the element to fouling requires frequent cleaning even in clean streams; this limits the practical application of this type of meter in commercial measurement and makes it useful mainly in research.

Another way to use thermal energy for measurement is to install two thermistors in a flow stream, one in the flowing stream and the second in a side pocket out of the flowing stream. The temperatures of both probes are kept constant, and a measure of the difference of power supplied is proportional to the flow rate.

A similar meter measures the temperature before and after a heat source with both probes in the flowing stream.

All of these meters are so sensitive to problems of dirty streams that they find application mainly in determining low velocity mass flow rate of clean fluids.

Tracer meters have been around for many years. The first units injected a foreign substance into a flowing stream and then picked up its presence at one of two detectors downstream of the injection. One of the first such meters injected salt into the water. The details of injection, detection, dispersion, flow profile, distance between probes, and marker types have resulted in many different meters and makeups. They are used for spot checks of velocity and, with an area factor, allow volume calculation.

A marker must have some characteristic that sets it apart from the flowing stream, such as the salt in fresh water, radiation, heat, and dye properties. The tracer should be approximately the same density as the flowing stream. It should mix well and travel at the same speed as the carrier. It should be readily available at low cost, chemically inert, and not naturally exist as part of the stream. It should be detectable by some standard analysis technique such as conductivity, color, radioactivity, chromatographic analysis, flame ionization, photometry, or heat sensing.

Most tracers operate intermittently by manual injection, but some meters measure flow continually with automatic injection based on time or marker detection. Injection and detection points are based on distance and dispersion of the sample matched against detector sensitivity. The most accurate measurement with these devices requires calibration in place, but they can be used with less accuracy for lines already in place by adding the injector and detector(s). They are quite often used in control measurement to set flow rates where the absolute value of flow is not critical.

References

American Gas Association. Measurement of Fuel Gas by Turbine Meter. Arlington, VA: AGA.

American National Standards Institute/American Society of Mechanical Engineers. Steam Turbines In: Performance Test Codes 6. New York: ANSI/ASME.

American Petroleum Institute. Chapter 5, Metering, In: Manual of Petroleum Measurement Standards. Washington, DC: API.

American Society of Mechanical Engineers. Fluid Meters, Their Theory and Application. 6th Ed., New York: ASME.

International Organization for Standardization. ISO 9951, The Measurement of Gas by Turbine Meters. Geneva, Switzerland.

International Organization for Standardization. Part 1, Orifice Plates, Nozzles and Venturi Tubes Inserted in Circular Cross-Section Conduits Running Full, In: ISO 5167, Measurement of Fluid Flow By Means of Pressure Differential Devices. Geneva, Switzerland.

13

READOUTS AND RELATED DEVICES

Secondary systems are a part of any measurement installation for handling primary element signals and the variables necessary to correct flow from flowing to base conditions. These elements fall into three main categories: mechanical, pneumatic, and electronic. All have applications in flow measurement. The choice can depend on a number of parameters, not least of which may be personal preference based on a given industry or company's experience. The fastest growing segments are the electronic systems designed to take advantage of the rapidly growing availability and value of computers.

Electronics

Several stages of development have taken place in the move to electronics. First was the glamour of being "up to date." Using the new electronics at that point often created about as many problems as were solved. The common reaction was, "electronic equipment does not work," because of the amount of work required to keep it running. Designers then came up with developments and improved capabilities, and users began a more studied evaluation of true needs and uses. These evaluations defined the most useful areas in which to apply the devices.

The present generation of transducers and computers is well received by users, more and more of whom are converting to electronics each year. There are several reasons for this user acceptance. The operating service (uptime) provided is equal to, or better than, mechanical and pneumatic types. Many additional capabilities are possible (e.g., smart transducers, etc.). Minimal maintenance is required. Trained personnel are available to install and maintain the units, and power requirements have been reduced to the point that auxiliary power sources such as solar charged batteries may be used for remote locations (Figure 13-1).

Fluid Flow Measurement. ISBN: 978-0-12-409524-3

Figure 13-1 Typical temperature transducer. Devices such as this, along with computers and other equipment, are vital parts of a metering system.

"Simple" computers may calculate a flow rate and totalize flow for a meter. Or computers can be operating centers for measurement, control, and communications in complex multi-meter systems. Computers can develop the complete volume calculation and print appropriate hard copy or feed a central control or computer center with the complete accounting procedures. They can provide real-time operation and control information for metering systems.

Their primary limitation is the cost justification compared to alternative ways to achieve desired measurement over a given service life. Systems for individual meters may cost only several hundred dollars, whereas larger meter stations can run to over half a million dollars. Cost effectiveness usually boils down not to accuracy considerations, but rather to efficiency of solving equations and the true need for speed. This is required in pipeline measurement to keep the bulletin boards current, as required by the Federal Energy Regulatory Commission (FERC).

The manner in which equations are broken down in standards and references indicates that their values are independent

Figure 13-2 Typical orifice meter installation.

and are individually addressed. This is an inherited interpretation based on the capabilities of past equipment. As long as conditions are relatively stable, the use of averages does not introduce significant errors. For varying flow, however, the basic equation requires the variables to be interpreted on a continuing basis with the readout system's frequency being faster than the flow system change. Such changes can be quite rapid, and the frequency response requirement demands the use of an electronic system for accurate flow measurement. A system that does not change much over hours of operation, on the other hand, can provide measurement without requiring continual integration from electronics related equipment (Figure 13-2).

Related Devices

Most related devices have an accuracy in the general range of ±0.15 to ±1% *of full scale*. This makes it important to choose transducers with the right ranges; measurements should be in the upper two-thirds of device ranges. The higher the differential, the more accurate the meter reading, provided flow does not exceed the differential device's range. Many users, not recognizing the effects that auxiliary instrumentation can have on the accuracy obtained by a given meter, compromise flow accuracy accordingly. The overall accuracy obtained includes the inaccuracies of each of the auxiliary devices and how they are used, as well as the primary meter accuracy. Sometimes the auxiliary devices may control the accuracy of the results more than the basic meter. For example, with gases (particularly near their critical points), a change of one pound can represent several percent in flow. This says two things: 1) it would be better to measure at some other location where conditions are farther away from critical points; or 2) the accuracy required for pressure measurement must be increased several-fold to maintain the same limits for the corrected flow measurement.

1 Online chromatograph
2 EFM
3 Modem
4 Battery (solar charged) EFM
5 Modem
6 Gas turbine meter
7 Gas turbine meter
8 Gas turbine meter
9 Flow

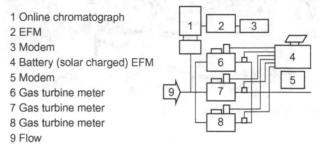

Figure 13-3 Typical, multi-unit flow computer system with three turbine meters.

Pressure measurement for liquid flow is straightforward. Liquids are generally less sensitive to their pressure measurement. However, in areas near the liquid critical point, density changes significantly, and pressure does affect flow measurement accuracy.

Location of the pressure tap for a meter is based on the meter calibration, in the same manner as one of the differential pressure taps for an orifice flow meter with gas flow. The tap into the line should be at a point specified so that flow past the properly made tap creates no undue turbulence, which can affect the reading.

The ideal point to measure pressure (at the point of velocity or volume determination) is usually not possible—or at least not practical—with most meters. When such mechanical problems make it impossible to install a pressure measurement tap at the proper point, then corrections may be required to account for the difference between the correct point and the actual point. Sometimes the difference is simply ignored if the difference does not affect density or flow calculations "seriously." The orifice equation for gas has an expansion factor in it to make the required correction. The difference can also be accounted for as part of basic meter-system calibration.

Pressure transducers must be calibrated on a routine basis to maintain accurate measurement. The standard most commonly used is a dead-weight tester or precision gauge for higher pressures, and a manometer for lower pressures.

Temperature transducers present problems that are similar to pressure transducers. Since actual volume flow normally is not at base conditions, a measurement of the flowing temperature is required to correct for the difference. Accurate measurement of temperature is more difficult than it appears since the transducers normally require insertion into the flowing stream and thus disturb smooth flow, which consequently disturbs a meter's operation if it is improperly installed. Therefore, temperature is normally measured at some point downstream of the meter after making sure that temperature will be essentially the same as the temperature in the meter.

The effect of ambient conditions on the readout equipment is also important. Radiation from the sun and conduction from pipeline heat can affect temperature readings by changing the temperature of thermowells and/or affecting transducer mechanics. For utmost accuracy, the instrument environment should be controlled by heating and cooling, shading, or insulation, depending on the required flow accuracy and the effect that temperature has on the fluid properties and flow measurement. Fluids near critical points are prime candidates for this treatment, whereas other less critical fluids generally require no unusual treatment. Smart transducers have helped minimize some of these ambient effects at less overall cost than some of the other treatments possible.

Differential pressure is the most important variable for differential head meters, since most errors in flow measurement with differential meters come from this measurement. These errors are so critical because differential pressure is the major variable in calculating flow. The maximum differential used with these meters is in the range of several hundred inches of water (i.e., less than a 10 psi drop). Quite often, static pressures may run hundreds of pounds, so the measuring device has a static pressure load on it of about 1,000 psia, yet it is trying to measure a difference in pressure of 1% or less of its static range. This requires the differential device to be rugged enough to withstand the high pressure requirement, yet sensitive enough to measure a very small pressure difference.

To minimize the differential problems (consistent with the required flow range), operate with the differential at higher values—provided the strength limit of the pressure-drop creator is not exceeded. Because of the cost of lost pressure, the differential pressure devices commonly used are in the 100 to 200 maximum of inches-of-water range. The devices used include manometers (used at low static pressures), diaphragm bellows,

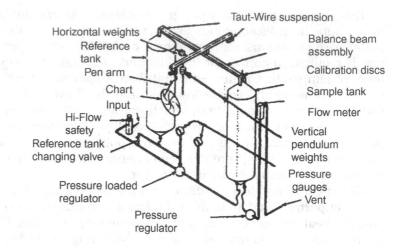

Figure 13-4 Air dead-weight used in testing differential devices.

and mercury-filled meters. These devices must be calibrated against manometers or dead weights. They are usually tested and calibrated at atmospheric pressure and then re-zeroed at line pressures. Some test devices that operate at line pressures are available, but their use is often restricted to laboratory work rather than field calibration.

Maintenance of differential devices consists of periodic calibration and, if necessary, replacement of driving mechanisms. In dirty service, periodic cleaning may be necessary, or the use of seal pots or isolating diaphragms may be required to prevent contamination (Figure 13-4).

Where wide ranges of flows are expected, multiple transducers can be used on a single meter to expand its range. For example, a more accurate low differential device, such as a 20-inches-of-water unit, can be manifolded into the same meter as a 200 inch unit. This combination expands the range of flow from 3 to 1 (on a 100 inch unit) to approximately 10 to 1 at similar accuracies. If ranges beyond this are required, then a second or third meter with proper valving can be used, with meters being switched in and out as the flow varies. Combinations of this sort allow an almost infinite flow range to be handled.

As previously stated, for the most precise flow measurement, the use of the smart differential devices is an investment of value to minimize calibrations and the effects of ambient conditions on the device.

However, there is a misconception that if a smart differential transmitter is statically calibrated down to less than 1 inch of water that very low flow can be measured accurately. This is an incorrect assumption. The statically calibrated smart transmitter does not operate in a static measurement environment. The differential head meter measurement environment is dynamic and as such experiences a noise element riding on the differential pressure signal. When the magnitude of the noise element equals the magnitude of the differential pressure that one is attempting to measure, differential pressure measurement error occurs. It is not a matter of a dumb Dp transmitter versus a smart Dp transmitter, it is because the Dp transmitter, any Dp transmitter, does not know the difference between noise and true signal. In addition, very low differential pressures in a differential head measurement mean very low velocities, which have been shown to result in ambient temperature influence, which in turn, biases the flowing temperature measurement and the flow by 0.1% per °F bias.

Relative Density or Specific Gravity

Reducing fluid measurements from flowing conditions to base conditions requires identification of fluid composition. The most useful parameter for this is the specific gravity or relative density of the fluid. Correlations in the petroleum industry are based on these measurements, and data for other mixtures are expressed in these terms. For pure products, the need for specific gravity reduces itself to the ratio of specific gravity at flowing conditions to that at base conditions, which allows for the correction of the effects of pressure and temperature on the pure product. Quite often a formula (equation of state) is available which expresses the effects of pressure and/or temperature, and specific gravity correction can be made with suitable measurements.

The several definitions of specific gravity used in the flow measurement business are important to understand. For natural gas, the definitions in AGA-3 are the molecular weight per unit volume of gas compared to the molecular weight per unit volume of air at the same conditions of pressure and temperature. This definition of "ideal specific gravity" ignores the corrections for compressibility when these measurements are made at atmospheric pressures, since such corrections are relatively small. However, this yields a specific gravity that is different by a small amount from the ratio of molecular weights including compressibility (which is equal to the real specific gravity). In non-natural-gas measurements, these definitions are not used. And the normal definition used outside of the natural gas industry for specific gravity is the ideal AGA-3 definition (i.e., the ratio of molecular weights).

In liquid measurement with the English system of units in the United States, the definition of relative density (specific gravity) is different, in that the weight per unit volume of the liquid is compared to the weight per unit volume of *water at 60°F*. Water at 60°F has a defined weight set by the International Steam Tables, so that a liquid specific gravity is directly convertible to density by multiplying the weight of water at 60°F times the specific gravity of the fluid.

$$\rho_f = (SG)(W_w) \tag{13.1}$$

where:

ρ_f = flowing density (or specific weight);
SG = specific gravity;
W_w = weight of water at 60°F.

This calculation is not possible with natural gas since there is no specification for the base air conditions, and hence no specific weight may be assigned (Figures 13-5, 13-6).

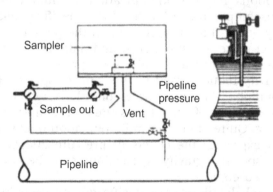

Figure 13-5 Recording gas gravitometer using the indirect weighing method.

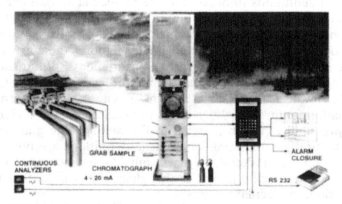

Figure 13-6 Typical gas sampling system.

When specific gravity alone does not sufficiently represent composition for flow calculations, then an analysis is required. This can happen when variable components make up a sample with the same specific gravity. Natural gases and mixed petroleum liquids exemplify the problem.

Sometimes there is a need to know the constituent makeup for pricing information if each component has a separate value. Corrections may be made for non-hydrocarbon constituents in the streams.

Sampling is a science unto itself, and great care must be taken to get a *representative* portion of the flowing stream for testing. Samples should be taken from sample probes installed in the lines extending away from the pipe walls into an area where good turbulence exists. A homogeneous mix should be present at this sample point. Getting a sample into a container, transporting it, and transferring it to the chromatograph offer many chances for introducing errors by distorting the sample characteristics. For this reason, in-line chromatographs with short sample lines running directly to the unit are typically used where practical.

Fluids difficult to sample include: light hydrocarbon liquids, gases at atmospheric pressure and ambient temperature, saturated gases, water and/or hydrocarbons, gases containing hydrogen sulfide, condensing gases or vaporizing liquids, crude oils containing water, and emulsions. In these cases, special procedures and equipment are required for sampling. But even when these considerations are recognized, getting good samples requires perseverance and often some luck. In the most sensitive cases, direct sampling into the analysis equipment is required. It is no field for an amateur to enter. For example, analysis values involved in the petroleum industry affect the exchange of money—and purchasers do not care to pay for crude oil when water has been measured as crude oil because of improper sampling. Likewise, natural gas improperly sampled may result in an inaccurate energy calculation—so the money is paid for a product not actually being delivered.

Crude Oil Sampling

A source of inaccurate measurement of crude oil is the inability to obtain a correct sample of a mixture of water and oil. Since "water and oil do not mix," special provisions must be made to mix the two liquids prior to sampling. This can be done in several ways:
- Normal mixing due to flow disturbance following a piping fitting such as an elbow in horizontal or vertical piping with sufficient velocity to cause the oil and water to mix.

- Static mixing with mechanical mixers (turbulence creators) inserted in the pipeline immediately upstream of the sample point. (Note: This mixer must be installed outside of the required meter tube lengths specified in the meter standards and is best located downstream of the meter.)
- Power mixing with an external pump that removes a small stream and reinjects it under pressure immediately upstream of the sample point. (Note: The mixing begins to reseparate within a few pipe diameters after the injection point.)

The choice of which method to use depends on the velocity in the pipeline and the piping configuration. The American Petroleum Institute (API) Chapter 8 presents complete recommendations for sampling liquids. The general requirements are for the flows to be well mixed, and that can be accomplished in several ways as shown:

It is critical to follow recommended procedures when sampling pipeline quality crude where water contents are low (below 1%). In large pipeline meter stations or ship loading installations with high flow rates, this small percentage represents a significant amount of money to both parties. The same procedures must be followed to ensure mixing at production meter stations with higher water contents.

Natural Gas Sampling

Many different methods have been approved and used in the gas industry to sample natural gas with varying success. Recognizing this, the industry—with the support of the Gas Processors Association (GPA), the API, the Gas Research Institute (GRI) and the Mineral Management Services (MMS)—sponsored testing at the Southwest Research Institute (SWRI) to completely evaluate the gas sample question (Table 13-1). The sampling results are used as indirect and direct multipliers for volume and determination of heat value. The inclusive tests and related work were completed and published last in 2006 as API MPMS Chapter 14, Section 1, "Collecting and Handling of Natural Gas Sample for Custody Transfer." This information should be carefully reviewed by the flow measurement practitioner in order to have the latest quality information and procedures for gas sampling.

Table 13-1 Suggested Minimum Velocities versus Mixing Elements

Mixing Element	Piping	Minimum Pipeline Velocity (feet per second)								
		0	1	2	3	4	5	6	7	8
Power mixing	Horizontal or vertical	Adequate at any velocity								
Static mixing	Vertical	Stratified	Not predictable	Adequately dispersed						
Static mixing	Horizontal	Stratified		Not predictable	Adequately dispersed					
Piping elements	Vertical	Stratified			Not predictable	Adequately dispersed				
Piping elements	Horizontal	Stratified					Not predictable		Adequately dispersed	
None	Horizontal or vertical	Stratified or not predictable								
		0	.305	.61	.91	1.22	1.52	1.83	2.13	2.44
		Minimum pipeline velocity (meters per second)								

Some of the evaluations that were studied and made available are the three methods of sampling presently used and covered in the standards:

Spot sampling;

Automatic composite sampling systems; and

Sampling into an online chromatograph.

These studies have been subdivided into these spot sampling areas:

1. equipment used;
2. physics of the gas to be sampled;
3. requirement of maintaining the sample temperature above dew points;
4. effects of sample-system material that comes into contact with the sample;
5. effects of the type of gas to be sampled on the choice of the process;
6. cleaning required; and
7. care of the sample after it is obtained.

This document covers sampling flows from those as simple as methane-rich streams (lean gas) that have been dehydrated and cleaned, to those as complex as samples as saturated gas flows. The value of this work is the empirical data that were taken to support the decisions made—some of which may require totally new and different approaches.

The report shows that root causes of gas sample distortions were found to include poor sampling systems and procedures, surface effects, thermodynamic problems, purge problems, and leaks.

From the document's results, it is clear that to obtain undistorted gas samples or to minimize distortion, attention should be paid to equipment cleanliness (particularly reused equipment), temperature of the sample equipment surfaces, hydrocarbon and water dew points, temperature at the sampled gas, the flowing thermodynamic state of the gas, (temperature, pressure, and composition) and the flow path of the gas (i.e., restrictions of the flowing cross-sectional area). The appropriate method for a particular flow condition should be chosen based on all of the above.

Some of the results indicate that present sampling systems may over- or underestimate the total heat value and density of the gas. The basic problem being addressed is getting the answer correct. Most exchange of natural gas is now based on total heat content, not just volume, as it was in years past. So the control of system economics is based on total heat flows within the system, not just a volume balance.

Analysis allows calculation of parameters important to flow measurement, such as relative density (specific gravity), heating

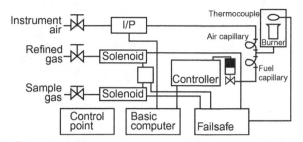

Figure 13-7 Schematic of principal elements of Thomas-type calorimeter.

value, compressibility factor, inert content, and density. Calculations are based on mixture laws and are accurate at base conditions, but conversion to flowing conditions is not easy and can, in certain circumstances, introduce errors where the mixture laws break down because of shrinkage (such as mixtures of light hydrocarbon liquids or two-phase gas flows).

The most common analysis instrument is the chromatograph. Based on standardized samples, chromatographs can be calibrated to cover wide ranges of fluids. Easily maintained, their calibration can be checked with a standard sample with a similar component makeup. The units come in models that can be applied continuously or intermittently when a sample is available. Most are permanently installed, but portable units can be used as line monitors at strategic locations until a problem arises elsewhere; then the portable unit can be taken to the problem site for on-the-spot analysis (Figure 13-7).

Calorimetry

Where a heating value is needed, a calorimeter can be installed to continuously monitor a stream. Alternatively, samples can be taken at meter locations and individual samples tested at a centrally located calorimeter. After all inputs have entered a pipeline, a single unit is often used to determine the heating value at all downstream locations. The choice between chromatograph and calorimeter depends on product value (quantity and cost) and the contractual requirement for corrections (i.e., most require correction for heat value, while some only require that a minimum heat value be maintained).

References

American Gas Association, Orifice Metering of Natural Gas and Other Hydrocarbon Fluids. AGA, Arlington, VA.

14

PROVING SYSTEMS

The necessity of proving a meter depends on the value of accurate measurement for the product being handled (Figure 14-1). Large volumes and/or high-value products and/or a need to reduce an unacceptable system balance are the prime candidates for using provers. Oil industry measurement of crude oil and refined products are examples of where meters typically involve proving systems. The proving systems are considered part of the cost of the meter stations and are permanently installed at larger facilities. When product value is lower, provers are usually portable (used within a limited geographical area); as product value drops further, proving frequency is reduced, and for the lowest value products proving is not done at all.

In other industries, proving in place is seldom done; metering is assumed correct until a process goes out of control or a meter breaks down and requires repair or replacement. For meters such as the orifice type, calibration is accepted to be correct as long as the mechanical requirements of the meter's specifications are met. Some meters are "tested" by calibrating only the readout units, with no test or inspection of the primary device. This does not have the same value as a complete system examination or the use of a prover (Figure 14-2).

In summary, testing can be a very expensive and time-consuming procedure or it can be as simple as an external physical examination during a walk by. Obviously, the ability of these tests to prove a flow meter's accuracy varies from the best that can be done to a "test" that really has nothing to do with flow accuracy. One of the easiest meter proving methods is to check the operating meter against a master meter that has a pedigree of accuracy and is adjusted for any differences in pressure and temperature. Any indicated registration differences can be calibrated into the operating meter by use of a meter factor applied manually or through an electronic meter adjustment. Meter factor in this case is defined as:

$$Meter\ Factor = \frac{True\ Volume}{Indicated\ or\ Meter\ Volume} \quad (14.1)$$

Fluid Flow Measurement. ISBN: 978-0-12-409524-3

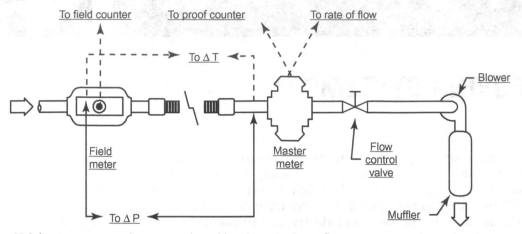

Figure 14-1 Low pressure proving setup using a blower to create gas flow.

Figure 14-2 Master liquid meter prover.

which "divides out" the meter indication and "multiplies in" the corrected volume.

Since most meters are not totally linear, tests should be run over the meter's operating range and the meter factor entered as an average factor over the range. Computers can apply a factor that varies with flow rate. The correction complexity required depends on the magnitude of meter non-linearity and the measurement's accuracy requirements. Proving is common for liquid meters, but is rare for gas meter systems.

Liquid Provers

For products whose vapor pressure is the same or less than atmospheric pressure under flowing conditions, an open tank prover may be used as a standard or field test measure. Standardizing groups can calibrate field test measures/cans (Seraphins) and stamp on them either the volume they contain or volume they deliver. Flow from a meter is diverted on the run from normal delivery to fill the can. Readings of the operating meter are taken at the start and at the completion of the filling. This procedure can be automated by the use of solenoid valves in the fill and bypass system.

At one time, similar systems for testing fluids with vapor pressures greater than atmospheric were used with closed containers. But the cumbersome field test measures/cans required have been replaced with the more convenient pipe provers. Pipe provers are available in several configurations such as standard and small volume, U-shaped, straight or folded, ball or piston displacer, and unidirectional and bidirectional. The choice depends on parameters of the job to be done (Figure 14-3).

The **bidirectional prover** requires a displacer round trip to complete one prover run. It can be made U-shaped, folded, or straight, depending on space requirements.

Figure 14-3 Liquid prover system.

The standard prover (U-shaped bidirectional) is the most common and uses an inflated ball displacer. Regardless of construction and operating details, all provers perform the same function.

When proving a volume meter, flow is passed through an operating meter into the prover. After temperature, pressure, and flow rate have been stabilized, the displacer is launched. For proving a mass meter, fluid composition must also be stabilized or any composition difference between the meter and the prover must be accounted for.

Since proving creates a temporary slowdown in flow until the displacer gets up to speed, some pre-run length in the prover must be allowed before displacement of the accurately measured volume begins. At a point after flow rate stabilization, a switch indicates entry of the displacer into the calibrated section, and the meter pulses are sent to the proving counter or circuit.

Flow continues until a sufficient number of pulses (typically 10,000) have been generated by the operating meter. An exit switch then indicates that the calibration volume has been achieved, and pulses to the proving counter are interrupted. Pulses generated by the operating meter are thus "gated" to the proving counter, without stopping the same pulses from going to the billing meter's counter. This displacer passage and collection of pulses is repeated a number of times (set by individual company policy but typically four or five), while the stabilized fluid pressure and temperature are recorded. Calculations convert the temperature and pressure to the same base conditions for the meter and the prover. When volumes are compared, the ratio of the prover to meter volume is the meter factor for this flow rate. Various provers have distinguishing characteristics (Figures 14-4, 14-5).

The **small volume prover**—sometimes called a "compact" or "ballistic" prover—has a precise pickup system that allows less tolerance in the switch location for the displacer. The timing system requires fewer actual meter-generated pulses, since they are "multiplied" to generate the necessary pulse rate to achieve 10,000 pulses between switches. To minimize the pre-run and flow interruption when the displacer is launched, the displacer is driven externally rather than taking its energy from the flowing stream. This allows stabilization to be reached rapidly with a short pre-run. This system allows proving with less volume displaced, hence the name "small volume" prover.

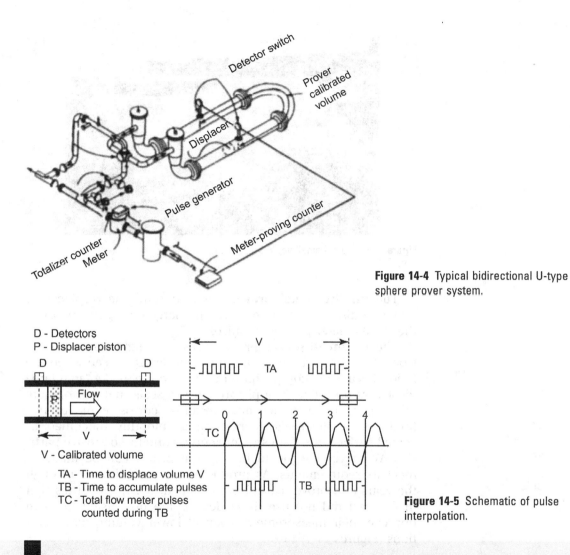

Figure 14-4 Typical bidirectional U-type sphere prover system.

D - Detectors
P - Displacer piston

V - Calibrated volume

TA - Time to displace volume V
TB - Time to accumulate pulses
TC - Total flow meter pulses
    counted during TB

Figure 14-5 Schematic of pulse interpolation.

Since these provers do not collect a minimum of 10,000 pulses, a special procedure known as double chronometry is employed, which captures a single pulse and designates it as representative of all pulses in the pulse train (i.e., assumes a uniform pulse train) and characterizes that pulse. Using that designated representative pulse's characteristics, proving electronics are capable of accurately counting a partial pulse, thus reducing the loss of pulses due to small volume.

However, if the meter does not produce a uniform pulse train then the captured pulse is not representative and the meter has difficulty achieving the required repeatability (Figure 14-6).

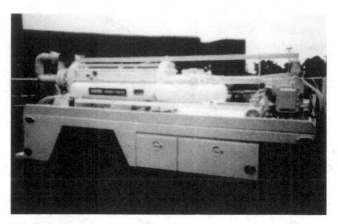

Figure 14-6 Typical small volume prover.

The **unidirectional prover**, built to send the displacer in only one direction, requires a volume large enough to produce the 10,000 meter-generated pulses.

The **piston displacer prover** uses a straight barrel design since the piston cannot go around a corner. It is used when the fluids, because of composition or temperature, make "standard" displacers unusable. The piston can use seals that will operate at temperature extremes and on most corrosive or reactive fluids. Since the seals operate on a smoothly machined or coated surface, the fluid stream should contain no erosive particles. Whatever the job to be done, a prover can be made to meet its requirements. As product value has climbed through the years and prover costs have dropped, many industries that in the past did not use these devices are now using them to improve their measurement, even at lower volume meter stations (Figures 14-7, 14-8).

Liquid Provers Summary

Proving systems are widely used in the petroleum industry as the basis for establishing flow meter accuracies as installed and operating on the operating fluid. A meter factor is determined and used until significant changes in flowing conditions occur.

Pipe provers are made in a number of different styles and shapes, and the choices are covered in API Standard Chapter 4 of the *Manual of Petroleum Measurement Standards* with details of their construction and use. Their proper use is the basis of accuracy in liquid flow measurement in the petroleum industry.

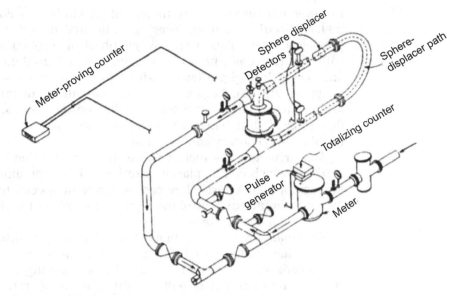

Figure 14-7 Typical unidirectional return-type prover system.

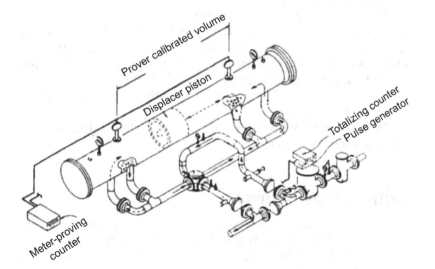

Figure 14-8 Typical bidirectional straight-type prover system.

Gas Provers

In the past, gas meters have not been proved like liquid meters. Proving an orifice meter has involved making sure that the meter's physical condition is maintained. In looking at ways to lower tolerances on gas meters of all kinds to reconfirm a meter or settle a concern over an individual meter's accuracy, proving is used. This occurs when physical inspection is not sufficient (such as with a PD or turbine meter), to define errors and actual throughput testing with a pipe prover is used. Some operators are also beginning to use adaptations of them with orifice meters. Other provers may be master meters, critical flow provers, or a centralized proving facility where meters can be taken for accuracy confirmation.

Master meters are meters whose basic calibration has been certified, which can be placed in series with an operating meter for a comparative test. They can be made in special test units with a computer to control the equipment, collect the data, and calculate a meter factor.

For small, low pressure meter testing, a low pressure blower can provide the test medium. The meter is taken out of service, depressurized, and piped in series with the proving unit downstream. The blower then pulls air through the operating meter and the standard meter to obtain proofs at a series of flow rates. A meter factor curve, plotted from these tests, allows an average factor to be obtained.

At larger volume stations with higher pressures, a master meter can be piped in series with the operating meter for a test. A computer again controls system operation, calculates the data, and collects meter factors. This system can be remotely controlled from a central office. Periodically the master meter is returned to a standards lab or to the manufacturer for recertification.

All of these systems have been successfully used to improve measurement accuracies.

Critical Flow Provers

One of the oldest testing devices for gas meters is the critical flow prover. The prover is installed in series with an operating meter that has been bypassed. Gas at operating pressure is passed through the meter, and then the critical flow prover, which is normally vented to atmosphere. If there is a nearby gas pipeline with pressure lower than the operating line by at least 15%, the gas can be passed into the second line without interfering with the test, and the gas is not lost.

Several differently sized critical flow nozzles can be installed in sequence at a test holder, or a variable pressure can be used for testing over a range of rates with a single nozzle. Meter factors thus determined may be used to correct readings of the operating meters, or to initiate a complete inspection of the meter to bring them into agreement.

The thermodynamic properties of the flowing gas must be known to calculate the flow at critical conditions. The usefulness of the method breaks down if the gas is near its critical temperature and pressure, where correcting factors are not adequately known. Likewise, if the gas contains condensed liquids or if liquids condense in the nozzle, the critical flow device cannot be used. Any deposits of solids from the flowing stream change the nozzle's throughput and must be removed before the nozzle is used for a test.

Critical flow nozzles are used to reconfirm meters (such as positive displacement and turbine meters) used in natural gas pipeline and distribution systems.

Pipe provers are similar to the small volume provers used in liquid testing; they have been developed and are used to determine gas meter accuracies (including those of orifice meters). The prover is piped in series with an operating meter, and a set volume is passed through the meter. Comparing prover volume to the indicated meter volume allows the factor to be determined. As the desire to reduce tolerances in measurement continues, this proving device is receiving additional attention and evaluation. Work to date indicates that, with careful testing and evaluating, a meter's operating accuracy can be achieved and a meter factor determined.

Central Test Facility

Where enough meters are in service to justify a significant test program, some companies employ a system of trading out meters (or meter internals) and bringing them to a central facility for testing on a periodic basis such as once a year. For example, offshore meters are often taken to an onshore facility for recertification. The centralized location where a standardized test facility is set up should have good quality gas flows available. The standard may be a master meter and/or a critical flow prover.

Gas Proving Summary

For the systems discussed above, economics must be evaluated carefully to determine justification limits for each method. These provings are typically done in response to governmental requirements or company policies where there is sufficient accuracy payoff to justify the investment required. This normally means high volume, standard natural gas situations or high price specialty gas systems. The American Gas Association's AGA Report No. 6, "Field Proving of Gas Meters Using Transfer Methods," provides guidance and calculations for all of the various natural gas proving techniques discussed.

References

American Petroleum Institute, Chapter 4, Proving. In: Manual of Petroleum Standards. API, Washington, DC.

American Gas Association, AGA Report No. 6, Field Proving of Gas Meters Using Transfer Methods, 1st Ed. Includes Errata.

15

"LOST AND UNACCOUNTED FOR" FLUIDS

Introduction

The report card for any business is the balance between what comes in, what is used by the business, and what goes out. This is referred to in the pipeline business as the "lost and unaccounted for" (LUAF) or system balance. In plant operation it is called the plant balance. In either case, the control of the cost of doing business and the profits earned are based on this report. It must be properly and continually monitored so that control is effected. Management must support the investment of time, money, and personnel if they want a meaningful report upon which to base business decisions.

Ideally, system balance/LUAF is managed using the laws of physical science. Physical science provides laws for conservation of mass and energy but not volume; not even volume normalized to a set of reference conditions. However, all too frequently, companies rely on contractual requirements for system balance/LUAF control. Contracts are an agreement between parties to provide a designated exchange (volume, energy, and/or mass) and they often rely on custody transfer measurement requirements for control. Custody transfer measurement requirements usually rely on industry standards that rarely address measurement operation, maintenance and volume/energy post-processing adjustment methodology. While industry standards and custody transfer requirements normally represent *good measurement practice*, they rarely provide sufficient controls for acceptable system balance/LUAF management. In addition, standards are focused on individual meter station exchanges. Thus combining multiple exchanges in order to form a system balance assessment can introduce differences if the exchanged material is not exactly the same, i.e., of the same relative density. Even though the standards attempt to account for these volumetric differences

through pressure, temperature, and compressibility normalization, they fail to address differences in relative density.

This issue is more significant for differential head meters where a difference of 0.001 relative density units can represent a 0.1% or more difference in normalized (standard or reference) volume. One might attempt to avoid this issue by conducting an energy balance or LUAF assessment, but total energy is not measured directly. The industry arrives at total energy, E, through the product of total normalized volumetric measurement, Q, and energy per unit volume (E/Q), i.e., $E = Q \times (E/Q)$, thus continuing to experience the issue. Contracts do not always account for normalized volume adjustments, i.e., transmitter calibrations, device corrections, physical properties corrections unless they exceed a stated percentage of average daily rate. Industry standards have been known to "grandfather" prior existing equipment or methodologies. This may be acceptable for custody transfer since it is an agreement between parties, but failing to incorporate all adjustments and/or "grandfathering" can build in either greater uncertainty (less control) or bias in the system balance/LUAF level. System balance/LUAF management needs to rely on the *best measurement practices, Physical Science Laws*, for controls. This is the only methodology able to minimize system balance/LUAF levels.

Controlling system balance is a matter of identifying influences that create differences between inlet measurement and outlet measurement and either eliminating them or reducing them to acceptable levels. Successful system balance control requires that it be properly managed. There is no basis for accountability if managing system balance or LUAF is not an identifiable function within an organization with dedicated responsibilities for monitoring, identifying, and reporting LUAF issues. The *best measurement practice* controls for system balance/LUAF management address all of the following:

1. Selection of the correct metering device for the application;
2. Installation of the selected metering device so that it may achieve its potential;
3. Operation of the selected metering system so that it may achieve its potential;
4. Processing of the metering system information so that it may retain its potential;
5. Maintenance of the metering system information so that it may retain its potential.

Pursuing sources of "lost and unaccounted for" should be a regular and ongoing process within a company dedicated to managing its system balance. (Auditing, to be discussed in

Chapter 17, is best done by company personnel not involved in running the system being audited, or by outside independent specialists.)

Much of what follows in this chapter has been discussed previously in this book. It is repeated here as it directly applies to "lost and unaccounted for" reports.

Liquid

Many of the factors affecting the uncertainty of liquid measurements in the oil and gas business are covered in detail in the API *Manual of Petroleum Measurement Standards* (MPMS). This contains standard procedures, equipment, terms, and petroleum fluid correction tables for the calculation of standard or net volume used in the transfer of petroleum liquids.

The MPMS is considered by many to be the most useful reference for liquid hydrocarbon custody transfer measurement. It represents the best industry practices. It is continually updated as more knowledge is gained. It is usually the criteria used in conducting liquid hydrocarbon measurement audits. Contracts refer to these standards, so when disputes or imbalances occur in the liquid measurement, the first check should be to make sure these requirements and practices are being followed.

Properly applied, the MPMS chapters ensure that both parties arrive at the same volumes and any disputes or losses are minimized.

Factors that can contribute to liquid volume differences include:

Losses to evaporation;

Leaks out of the system or within;

Bookkeeping or accounting errors;

Incorrect deliveries;

Thefts; and

Limitations of the equipment used in the system.

Evaporation losses and leaks are two unmeasured losses out of the system. They are estimated based on company or industry levels, but are usually *not* the major source of the balance problems.

Bookkeeping and accounting include all records feeding the measurement balance including:

Measurement tickets;

Calibration records;

Tank tables;

Logs and schedules;

Calculations; and

Data transfer.

All of these records should be reviewed for obvious errors or data not in line with the "normal" data for a specific station.

Full reviews of all stations, including design and installation, are required to ensure that all installed equipment conforms to the API standards.

Dynamic Metering

The meters most commonly used for dynamic measurement of petroleum liquids are the orifice, ultrasonic, turbine, displacement, and Coriolis types. They should be reviewed to ensure that they are the proper choice for a given station, based on the meter's characteristics, operational requirements, and the physical properties of the liquids. Operating the meters beyond their capabilities, or with their established meter factors, may cause problems. Such parameters as viscosity, temperature, and pressure change must be reviewed. Specific questions that must be answered are:

Are the flows staying within the range of the meter being used?

Are the meters capable of measuring the flowing parameters of temperature, viscosity, density, contaminants in the liquids, and corrosive liquids?

Are the meters protected from unstable flows by proper use of air eliminators, surge tanks, and relief valves?

For liquids close to their vaporization points, is sufficient back pressure being maintained, and are records available to check the settings? Is a flow-conditioning device properly installed and checked to ensure it is clean?

Static Metering

This is usually some type of tank gauging or vehicle weighing system where the following data must be checked and verified:

Liquid level readings;

Specific gravity;

Liquid temperature;

Free water;

Viscosity;

Strapping tables;

Tank cleanliness (incrustation on walls);

Foreign material in the system;

Tank tilt;

Dead wood;
Tank floor stability; and
Scale calibration.

Comparison of Dynamic and Static Metering

When two different types of measuring systems are used, the proponents of one type will question the capability of the other type. However, properly applied, both systems can achieve comparable uncertainties.

Tank measurement accuracy depends on the total volume in the tank. With a small volume in the tank, the percentage of error caused by the limitations of the level measurement is critical. Therefore, operations of the gauging, timing, and volumes present should be monitored carefully.

Dynamic meters should be proved on a regular basis. However, unacceptable changes in a meter's performance and/or a prover's function must be monitored by evaluating the magnitude and direction of the changes in the meter's meter factor under similar operating conditions. Meter factor control charts are recommended for monitoring the meter's performance and identifying any unacceptable changes.

Specific differences found to cause problems in the meter include:

Dynamic meters:
  Proving frequency versus flow parameter changes (pressure, temperature, and/or relative density);
  Maintenance done on meters and/or provers; and
  Operating condition changes.

Tank gauging:
  How long are liquids allowed to settle in the tanks?
  Are the tank strappings current?
  Is the water properly measured and corrected for?
  Stability of tank bottoms.

Summary of Liquid Balance Studies

Determining why volumes vary in liquid measurement requires the commitment of people, time, and money. Are inventories correct? Do the volumes of the terminals and the pipelines agree? If not, then a complete review of the documentation, the meter stations, and the operating and calculation procedures must be undertaken. Corrections of variations found will normally answer measurement questions and lead to an acceptable balance.

Gas

At one time, significant amounts of gas were actually lost through leaks in pipelines. This was many years ago when cast iron pipelines with bell and spigot-type joints were used for low pressure manufactured-gas distribution. Because of the porosity of the cast iron and the lack of sealing of bell and spigot joints, leakage was significant. Allowances were made for this gas as an unmeasured loss. It was a part of the economic calculation of the pipeline operating costs.

However, with the conversion to natural gas, involving higher-pressure, higher-volume production and long-distance pipelines, leaks were found by the presence of discolored vegetation or by spotting gas leaks—primarily made noticeable because of high-velocity noise—when walking the lines.

Other than very small leaks such as valves and flanges, pinholes, intentional blowdowns, and unexpected line losses, most unaccounted for gas today is caused by the limitations of flow measurement resulting from poor meter application, operation, and maintenance. Under the best of circumstances, all flow measurement has uncertainties that prevent 100% accuracy being achieved. Thus, the challenge of controlling the lost and unaccounted for gas is always present to some degree.

Where are these measurement uncertainties? The principal concern is to determine whether there is a problem of significance or whether the balance results fall within "realistic expectation." So what are realistic expectations? Two sources that help to define realistic expectations are: (1) an operating company's past history of specific measurement balances; and (2) the experiences of similar operating companies' balances.

Depending on system and flow measurement complexities, a realistic expectation can vary from ±0.25 to ±0.5% for large pipeline companies, and from ±3.0 to ±10% for production field balances. Distribution companies usually fall somewhere in the ±3.0 to ±20.0% limits. All unaccounted for gas is lost revenue, so there are economic reasons for finding the sources of the loss (considering costs versus savings). Estimated savings depend on identifying the sources of differences, which can be quite small and not always easily found.

The Meters and Fluids

As previously mentioned, a system balance review should include the type of meters installed, their location, their installation, and sizes and types including primary elements and associated readout equipment. Equally important are fluid properties and how they are being determined. From this information, an estimate of *expected* system uncertainty can be determined. The review should include operating ranges and maintenance history of each station with emphasis on the larger volume stations as potential sources of significant loss.

The meters themselves should be compared to the latest industry standards and procedures. Flow measurement is an evolving practice, particularly with new meters impacting system balances. Operational procedures and standards are updated continually as new knowledge is obtained on the various meters' (old or new) performance and the means of reducing uncertainties. An example is the 2000 revision to AGA Report No. 3, Part 2, which revised the installation requirements in order to reduce measurement uncertainty. This kind of knowledge can be used to evaluate previously installed meters to see if there are advantages to rebuilding and improving large-volume meter stations.

To verify meter selection and system operation, the following data should be collected and analyzed.

Meter

> Meters (number and location);
> Volumes measured at each (range and total);
> Measurement variable;
> Types;
> Readout system;
> Accuracies expected;
> Range;
> Station design and installation;
> Operating procedure;
> Maintenance procedure;
> Fluid condition;
> Calibration test reports; and
> Maintenance reports.

Information Flow

> Field (electronic or charts);
> Communication (procedure and controls);
> Office (procedure and controls); and
> Accuracy checks at each point.

Once a problem station is identified, complete examination of the station should be made including:

> Meter and meter installation to check for compliance with industry and/or company standards;

Gas quality meets contractual requirements with no carry-over of solids or liquids;

Inspection of meter tube, secondary (transducers) and tertiary equipment (computers) to confirm they meet standards followed by a thorough test report review; and

Maintenance procedures checked for recurring calibration problems that may require upgrading or change of the present equipment.

Other sources for review are the leakage determination programs and reports as well as possible theft.

Meter Data

Depending on the type of system employed (e.g., electronic or manual charts), flow information must be moved to a central office to complete the billing process. This handling and rehandling of data must be controlled at all points: in the field, the transmittal system, the office data system and the billing system. Checks and auditing, including integrity checks throughout the process, can confirm that information is being moved without degradation (Figure 15-1).

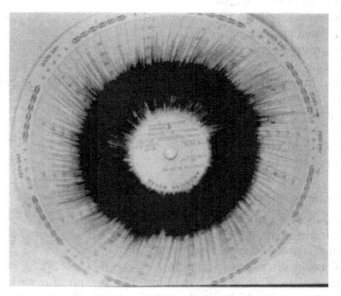

Figure 15-1 A chart like this presents major integration difficulties and can lead to large errors in flow measurement.

The Fluid

Fluid characteristics and their effects on meters must be reviewed. With production gases there are often problems in maintaining a single phase and getting a legitimate sample for determining the heat value and other properties. Pipeline quality gases that have been separated and dried do not have as many problems.

What is the Magnitude of Savings?

Some idea of savings can be obtained by evaluating one of the major factors—metering uncertainty.

A meter study should concentrate first on stations with higher volumes. Quite often 90% of the total flow through a system is being metered by as few as 10 to 20% of the meters.

A 0.1% additional uncertainty or bias on a station handling 500,000 Mcf/day is equal to *$20,000 per day* for gas that sells for $4 per Mcf. On the other hand, the same additional uncertainty on a volume of 50 Mcf/day is *$2*. The error in this reasoning occurs when there are enough 50 Mcf/day stations to equal a 500,000 Mcf/day station.

For larger additional uncertainties or biases—in the 5, 10 or 20% range—meters measuring smaller flows become proportionately more important. This approach allows a program to be planned that has the best chance of finding unaccounted for gas with sufficient economic significance.

The foregoing are the major sources of measurement differences that produce unaccounted for gas. These field problems have increased in recent years because:

Maintenance and testing (time and personnel) have been reduced with budget reductions.

Auditing of the whole measurement process has been reduced to a minimum, if it is conducted at all.

Management is reluctant to spend money unless they can be guaranteed a return on investment of some set percentage (e.g., 15 or 20%). Putting definitive numbers on flow measurement uncertainty is at best an estimate, but is usually less than these values.

A great deal of upper management's experience does not include operations, and therefore they are ill equipped to understand flow measurement problems and how to solve them.

Most unaccounted for gas will be "found" in field measurement problems. The solution to the problems will involve short-term expenditure, but will result in long-term income.

It is a bit daunting, however, to convince management who are not looking beyond the short-term profit and loss report to make a long-term investment of time and money.

Flow measurement solutions have been known to the gas industry for 50 years, but the gas industry has changed. The company that wants to find its unaccounted for gas should go back to the basics of good measurement. There is no silver bullet; finding LUAF losses requires a lot of hard work. But it is work that will pay off in terms of increased income.

Summary of Gas Balance Studies

The procedures outlined above are not a "one time and forget it" process, but should be instituted as a continuing operating procedure with all company personnel dedicated to minimizing the problem. Flow measurement problems found should be corrected as soon as defined. The data integrity from origin to final billing must be followed, analyzed, and audited in as close to real time as possible so that the quantities of gas measured are correct and current. This corrects for errors with a minimum of time required to correct billings, and it minimizes the effects on a company's profits.

16

MEASUREMENT DATA ANALYSIS

Introduction

During the course of performing the fluid measurement function, large quantities of field measurement information is gathered, either manually or electronically, in support of the quantity determination, contractual and/or regulatory compliance, and/or operational support. This information is in the form of uncorrected volumes, corrected volumes, mass, energy, energy per unit volume, fluid analysis, average differential pressure, average static pressure, average flowing temperature, flow time, average velocity, and, if available, meter diagnostics. This information is normally stored as hard copy or in an electronic data warehouse. The review of such data, if it occurs, is usually relegated to the overburdened field measurement technician or, in the office, to an overburdened individual or one who does not have the experience to assess the information effectively. Therefore very little benefit is obtained from the archived measurement information. In fact, the archived information is usually only reviewed when a question is raised concerning a specific metering point. However, if routine analysis is performed on the archived measurement information in the form of measurement control charts, a great deal of useful information can be obtained and used to determine the measurement health of a given metering point. Even more importantly, that analysis information can be used very effectively to manage system balance/LUAF levels. The following discussion will demonstrate the type of benefit that be obtained from routine measurement information analysis.

Corrected Volume

In all cases corrected volumes are required to be archived for a substantial period of time (years). Analysis of these volumes

Fluid Flow Measurement. ISBN: 978-0-12-409524-3

can provide insight into system balance/LUAF levels and direction. As an example on a system that is primarily composed of differential head meters and is already showing a significant (greater than 0.6%) annual negative loss per year, a corrected volume control chart that shows corrected volume input versus corrected volume output could indicate the outcome of increasing the volume through the system. Figure 16-1 is such a chart and it indicates that increasing the volume through the system will most likely increase the loss.

Knowing that the system was composed mainly of differential head meters led one to suspect that the system was suffering a serious contamination problem in the outlet meters. Differential head meters tend to under-register when contaminated. This was borne out through site audits of the 60 controlling outlet metering stations. Once identified and corrected, the system balance/LUAF returned to acceptable levels and the system was able to increase its volume.

Differential Pressure

Analysis of the archived average differential pressure can provide a number of indicators:

1. Meter station average low hourly or daily differential pressure—the differential head meter is most likely too large, see Figure 16-2. Best measurement practices recommend that the differential to continually operate above 10 inches of water or 10% of the upper range value of the transmitter.

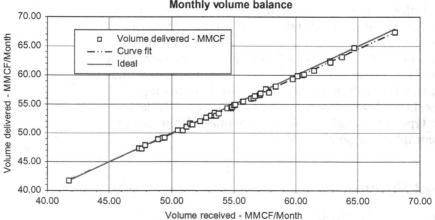

Figure 16-1 A comparison of system inlet corrected volume/year versus system outlet corrected volume/year.

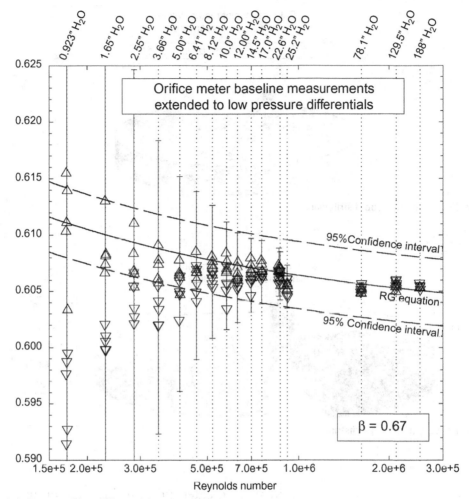

Figure 16-2 Influence of low differential pressure.

Occasional excursions below the recommendation are acceptable, but only for very short period of time.

Figure 16-2 was developed while flow-calibrating the coefficient of discharge, Cd, for an orifice plate using a smart differential pressure, Dp, transmitter which was statically calibrated to less than 1 inch of water column, iwc. During this calibration the Dp was selectively reduced to less than 1 iwc. In doing this the Cd data points show increasing data scatter until at less than 1 iwc the scatter is in excess of 5%. A flow computer sampling at the low Dp

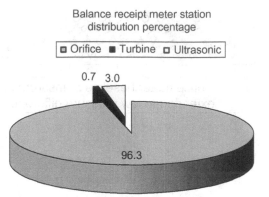

Figure 16-3 Receipt meter-type distribution.

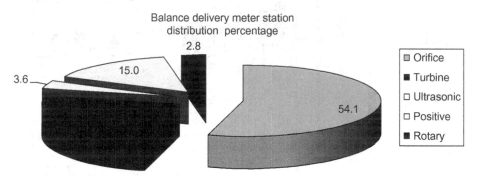

Figure 16-4 Delivery meter-type distribution.

would incur a greater flow uncertainty of in excess of ±2.5%.

2. System-wide seasonal low differential pressures—on a system that is composed predominately of differential head meters, see Figures 16-3 and 16-4, this indicates seasonal system load fluctuations.

From the pipeline system balance/LUAF shown in Figures 16-3 and 16-4, it can be seen that there appears to be an established seasonal relationship. This seasonal relationship shows that the system balance/LUAF is less in the winter but greater in the summer. This appears to be a load (throughput) relationship. Looking at the number of low monthly Dp receipt and delivery meters shows that the receipts do not emulate the seasonal system balance/LUAF relationship (Figures 16-5 to 16-8).

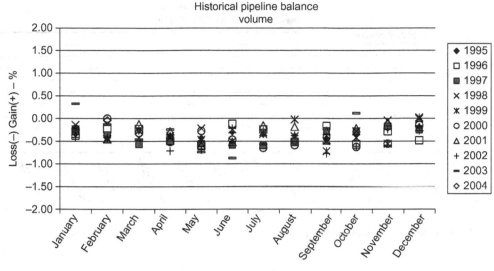

Figure 16-5 Percent monthly system balance.

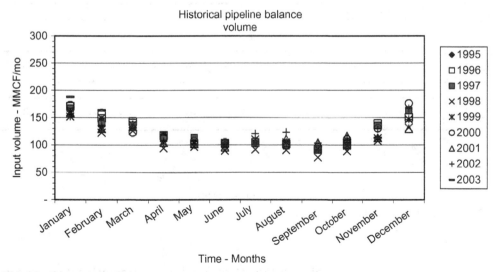

Figure 16-6 Monthly system volume.

However, the number of low monthly Dp delivery meters do emulate the seasonal system balance/LUAF relationship. The larger the number of low Dp meters, the greater the system imbalance/LUAF percentage loss. Therefore, the issue is within the delivery volumes. Pipeline personnel tend to

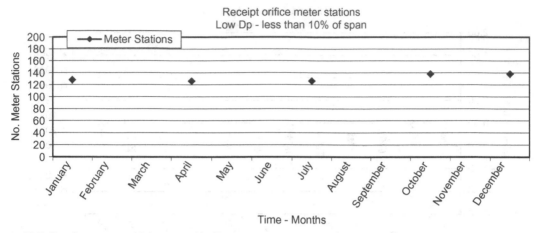

Figure 16-7 Receipt meters with low monthly Dp.

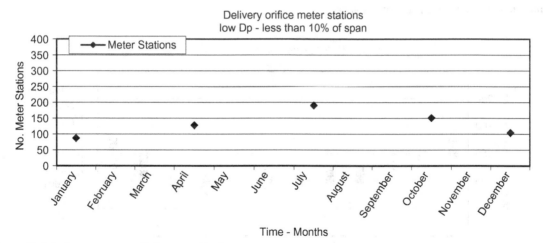

Figure 16-8 Delivery meters with low monthly Dp.

install the largest orifice bore diameter that their measurement procedures will allow in order to avoid being called out to change plates when deliveries increase, which results in low Dp during low delivery periods. This is a low Dp (less than 10 iwc) issue. Thus changing the orifice plate bore size to maintain higher Dp during low delivery summer months will improve the system balance/LUAF.

Flowing Temperature

Determination of the flowing temperature would appear to be a rather straightforward operation. A temperature measurement device is installed in the pipe in a temperature well (thermowell) in which a conducting fluid has been placed. The thermowell is used in order to isolate the temperature transmitter from the pipeline operating pressure. The output of the temperature transmitter is monitored and data from it is collected for onsite or remote determination of flow. The average flowing temperature is archived in support of the volume determination. This is a very simple concept on the surface, but it does have its complexities. In the process of performing the temperature measurement function, the temperature transmitter is exposed to the ambient environment of the pipeline, in particular the ambient temperature, resulting in conduction, convection, and radiant heat transfer with the ambient acting as an infinite source or sink. This ambient heat transfer tends to influence the thermowell and the pipe to which it is attached. This influence usually results in a biased temperature measurement which is a function of the ambient temperature. If the fluid entering the metering station is cooler than the ambient environment, then the resulting temperature measurement bias would be an over-registration of temperature. However, if the fluid temperature is warmer than the ambient environment, then the resulting temperature measurement bias would be an under-registration of temperature. In gas measurement, the measured temperature bias would be 0.1% per degree for differential head meters and 0.2% per degree for linear meters.

This issue is amplified by the size of the pipe and the velocity of the entering gas; the lower the gas velocity and the smaller the pipe, the greater the bias. In addition, covering the meter run or station will only mitigate the radiant component of the influence. Figures 16-9 and 16-10 demonstrate this influence on a differential head meter and a linear meter.

Figures 16-9 and 16-10 are for the same meter with the same ambient temperature environment but with different differential pressures and gas velocities. The gas temperature entering the meter run/station was 60°F in both cases. The influence of ambient temperature, as one might suspect, does not appear to be linear (Figures 16-11 and 16-12).

These data show that this influence is not a meter issue but rather a piping and velocity issue. It becomes a measurement issue when the "measured temperature" is used in the

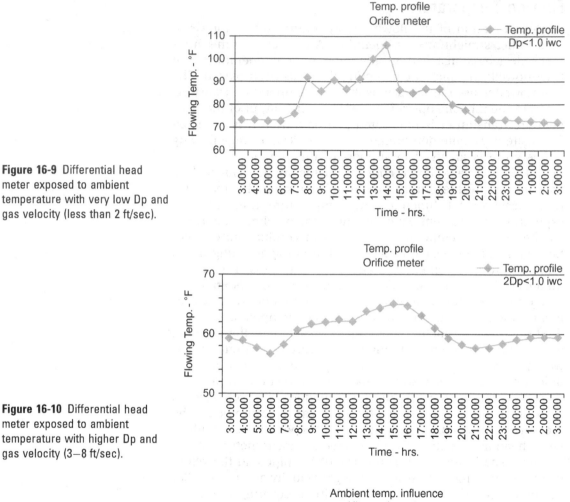

Figure 16-9 Differential head meter exposed to ambient temperature with very low Dp and gas velocity (less than 2 ft/sec).

Figure 16-10 Differential head meter exposed to ambient temperature with higher Dp and gas velocity (3–8 ft/sec).

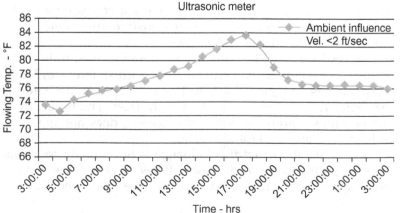

Figure 16-11 A covered linear meter exposed to ambient temperature with low gas velocity (less than 2 ft/sec).

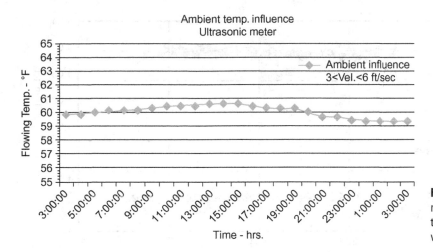

Figure 16-12 A covered linear meter exposed to ambient temperature with higher gas velocity (3—6 ft/sec).

calculation of corrected volume. If all the meters in a system were of the same type, differential or linear, and all experienced the same ambient temperature environment, then the influence would not significantly impact the system balance. However, this is not normally the case.

The liquid segment of the energy industry appears to be more conscious of this phenomenon, since it is common practice to insulate from the meter downstream through the temperature measurement thus mitigating this influence.

Fluid Analysis

The volumetric measurement of a fluid mixture, as opposed to a pure fluid, requires analysis of the fluid in order to identify the magnitude of the various components of the mixture. Knowledge of the components of the mixture is essential for volumetric measurement. One might attempt to avoid the need to analyze by utilizing direct (as opposed to inferred) mass measurement. However, this results in knowing the mass of a mixture which is usually not adequate. In natural gas volumetric and energy measurement, the compositional analysis is used to determine the energy per unit volume and the relative density (specific gravity). Maintaining control charts on the analyzed properties can provide warnings of incorrect values; see Figure 16-13.

Flow Time

Flow time in hours, days, and/or months can provide valuable information on the proper operation of the measurement

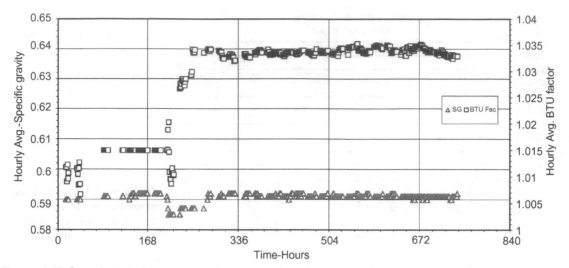

Figure 16-13 Control chart of energy per unit volume and relative density determination.

Time	PF_Psig	DP	TF	Flow time
3:00:00	0.11	283.44	73.4	49.55
4:00:00	0.09	283.31	73	27.28
5:00:00	0.06	282.23	72.9	2.33
6:00:00	0.08	279.61	72.6	17.48
7:00:00	0.03	276.03	76	0
8:00:00	0.11	274.23	91.3	14.48
9:00:00	0.39	275.33	85.5	60
10:00:00	0.38	274.52	90.5	60
11:00:00	0.53	275.8	86.5	60
12:00:00	0.38	275.27	90.8	60
13:00:00	0.11	275.17	100.1	57.45
14:00:00	0.2	274.5	105.9	50.83
15:00:00	0.28	275.88	86.1	59.93
16:00:00	0.05	276.44	84.8	9.07
17:00:00	0.03	275.82	86.7	0
18:00:00	0.1	274.07	86.9	20.43
19:00:00	0.39	275.37	79.5	60
20:00:00	0.21	274.5	77.3	60
21:00:00	0.26	275.61	73.5	24.98
22:00:00	0.22	275.29	73.3	48.25
23:00:00	0.08	277.44	73.2	0.02

Figure 16-14 Control chart of flow time.

station. Successful completion of the designated flow time period assures the practitioner that the measurement has remained within the acceptable range of the meter, thus allowing the practitioner to assess and/or manage the overall measurement uncertainty. Figure 16-14 shows a metering installation where the flow time consistently fails to complete the time interval, hour. This may be an indication of an oversized meter, in the case of a linear meter, or the need to adjust the bore size of a differential head meter, i.e., an orifice meter. It can also indicate that the metering range is so large that an additional meter or a different type of meter is required.

Consistently failing to complete the time period usually results in under-registration of flow.

17

AUDITING

Introduction

Information in this chapter should be supplemented by reference to other chapters for a complete understanding of the advantages and limitations of individual meters.

Physical auditing, as opposed to financial auditing, is a formal periodic examination and check of the flow measurement equipment and procedures of a specific company and/or location. It should include field and office operations from the measurement source to the end user, including data reports, overall performance, field operation, data handling, calculation, accounting, and final billing.

The client for which an audit is performed will be best served by using personnel to conduct the audit who are *not* directly concerned with the flow measurement system being audited, or by hiring outside independent specialists with a proven track record rather than performing the audit in-house. The outside viewpoint has no axe to grind and will usually provide a more thorough and valid report. Also, the specialists in this field bring a wealth of experience and knowledge, and they know where to concentrate efforts for the most benefit to the client.

When money is exchanged for measured fluids, the agreement will usually include a means of auditing the quantities obtained. Sufficient operation and maintenance records should be made available to all parties so that the calculated quantities can be determined independently. At least a check of the values used by the other party should be made to see that agreement is reached on the calculated quantities.

This procedure is an important aspect of custody transfer metering and is usually completed within 30 to 60 days after the required data is processed. It will keep both parties involved and prevent disagreements about procedures and quantities at some later date. With the data still current, any disagreements can be settled while knowledge of the measurement is fresh in the minds of both parties.

Fluid Flow Measurement. ISBN: 978-0-12-409524-3

A complete record of all disagreements must be maintained, including their resolutions. Records can be reviewed to see if a particular meter station or particular differences are recurring problems that need to be addressed by an equipment or maintenance upgrade. Most flow metering is done with the help, if not the specific use, of computers. This, therefore, is where much of the information can be found. However, a meaningful audit should go beyond just data review and include some analysis of the data's quality.

The following is a general overview of items related to meters typically included in gas and liquid audits.

Gas Meters

Number, type of meters, size of meters, any manufacturer's calibration reports on meters for all gas meters used for accounting purposes;

Installation configuration and dimensions for all meters;

How often are meters and instrumentation calibrated/tested?

What do typical flows look like? (maximum? minimum? average?)

How is the composition of the gas determined?

What procedures are followed to check meters, transmitters/recorders and flow computers?

Copies of meter test reports, volume data, and gas analysis data for the audit period;

Verification of all flow computer configurations and volume/MMBTU computations; and

Witness meter test and calibration.

Liquid Meters

Provide certification data on all liquid provers;

Provide certification data on all pycnometers (calibration cells);

Number, type of meters, size of meters, any manufacturer's calibration reports on meters for all liquid meters used for accounting purposes;

Installation configuration and dimensions for all meters;

Is there a kinetic mixer or some other mixer type to make sure a representative sample of the liquid is analyzed?

How often are meters, densitometers, and instrumentation calibrated/tested?

What do typical flows look like? (maximum? minimum? average?)

How is the composition of the measured liquid streams determined?

What procedures are followed to check meters, transmitters (recorders) and flow computers?

Provide copies of liquid meter run tickets, liquid volume data, liquid analysis data, and meter proving reports for the audit period;

Provide copies of pycnometer (calibration cell) calibration report for the audit period;

Give an explanation of how all meter and densitometers factors are applied;

Verification of all flow computer configurations and volume/MMBTU computations; and

Witness liquid meter test and calibration.

Analysis Equipment

How many onsite and lab gas chromatographs are in use?

How are they certified?

How often are they calibrated?

What calibration procedures are used?

If spot samples are utilized for accounting purposes, what procedures are used to obtain the samples, and what is the purpose of the spot sample?

If sample bottles are used for accounting purposes, what is the cleaning procedure?

Audit Principles

Flow measurement is the cash register of our business, and minimum uncertainty directly relates to profit control. The principles of an audit are the same for gas or liquid operations (as outlined by the Institute of Internal Auditor procedures) and the audit should:

Define the object of the audit;

Design the audit procedures;

Collect the necessary audit evidence;

Perform the necessary tests to confirm the uncertainty of the data;

Form an audit conclusion; and

Institute corrective action on the problems found.

The specifics of a gas flow and liquid flow measurement audit are quite similar, except that gas audits may involve chart integration, manual or electronic, and liquid audits include run or batch tickets.

Objective

The audit's objective is a broad description of what is intended to be accomplished. It should address the risk associated with a specific activity. It should be stated clearly, so that procedures, evidence, and testing are clearly outlined. A clear objective helps other audit participants to understand the scope and to determine the level of liability. It sets limits for the work to be done.

The scope of a measurement audit can include confirming and proving any or all of these areas:

Compliance with the contracts and company, industry, and/or government policies, procedures, laws and regulations;

Gathering information and verifying that it is correct, reliable, timely and complete—and properly used in the financial transactions;

Accomplishing the audit in a businesslike manner without waste;

Protecting company assets; and

Assuring that the audit is carried out in a way that does not interfere with the organization accomplishing its operational requirements.

The level of risk that management is willing to accept will determine the depth of the audit and the significance of deviations plus the need for correction.

Procedures

Audit procedures prove and test how well internal controls are working. Procedures detail how each testing is required to meet the audit objective given the internal controls that are in place. Proper internal control should be verified. However, this condition should be proven with minimum audit procedures or definitive testing.

Evidence

An audit of flow measurement requires collecting, analyzing, interpreting, and documenting sufficient useful information to

support audit results. If information is available in computerized form, data collection is simplified.

Audit evidence forms the basis of audit conclusions, opi-
nions, findings, and recommendations. Evidence should be correct and relevant to the objective. It should be complete enough
to ensure that another auditor would come to the same conclusion. Evidence includes:

Analytical evidence: review of sets of data (statistical quality
control charts, vessel correction factor data, meter factor
control charts, etc.);

Meter station evidence: information created about the
audited process and stored in a permanent form (meter station installation and operating data, meter tickets, calibration reports, invoices);

Physical evidence: information from observation or inspection of the activity, test personnel or meter stations. This evidence is documented from field reports signed by both
parties to the contract (observing meter provings or tank
gauging, meter station drawings);

Testimonial evidence: interviews or written statements from
meter personnel who are familiar with the activity under
review (technician's description of procedures by which
meters are calibrated).

Any form of evidence can be used to meet the audit objective. However, each must be applied with an understanding of
its limitations and weighted accordingly. While testimonial evidence is relevant, it must be corroborated by both parties,
because the possibility of bias can make accuracy questionable.
Physical evidence is the most basic, but gives no information
about ownership or value. Also, information may be biased as
people defend their behavior. Documentary evidence is a report
of test work done. It is one step removed from the actual test
event or activity. Money is exchanged on the basis of the document, and it should be checked by witnessing actual tests.
Analytical evidence can point to the need for further checking
before drawing conclusions.

Definitive Testing

Definitive testing consists of detailed testing and analytical
procedures designed to detect quantity misstatements or significant control weaknesses. There are two kinds of statistical testing; quality and variable. Quality testing ascertains the
probability that a specific condition does exist. It is a "yes/no"

test and is useful for determining whether internal controls are in place and working. Variable testing allows auditors to make probability statements about values and amounts. It is used to test quantities—volumes, dollars, time, and similar variables.

When the test population is large, statistical sampling is usually employed. This allows the auditor to:

Generalize test results for the entire data set; and

Specify audit risk mathematically.

After evidence has been gathered and definitive testing is completed, the auditor, using the facts and professional judgment, forms an audit conclusion. If the auditor concludes that significant deviations occur in tested areas, he or she will write a detail report. A detail report includes a condition statement, criteria, root cause, risk, and a response from the entity audited. Background information, findings, and recommendations for corrective action should be included.

In reviewing flow measurement procedures, the auditor provides assurance that trained personnel are measuring the product and following proper procedures with appropriate equipment and control.

In any of the audits listed, one of the first decisions to be made is to determine whether all stations should be audited, or if it would be more cost effective to select only a smaller group of stations. Because of limited time, manpower, and costs involved, not all meter stations are audited often. Each company should choose the stations to be included. The type of station (receipt, transportation, or delivery) may not be as important as evaluating:

Volume handled by the station;

Metering by others;

Exchange points;

Known problem stations;

Inspection report review; and

Other company questions.

The majority of pertinent data may be computerized, with electronic equipment performing calculations from electronic transducers plus data transmission and storage. Internal audits should be set up to compare field information from site inspections to the stored data used to perform calculations. Calculations may include equipment verification (plate changes, transducer calibrations, etc.), meter and peripherals testing frequencies, test results, lost and unaccounted reports, and contract requirements.

Sources of Information

Auditors depend on other people (field technicians, outside companies, other departments), copies of paperwork, charts and computer-generated reports for the necessary data. None of these sources is completely error free, so it is important to compile all required documents before beginning the audit. Separate files should be set up for each meter station chosen for the audit.

What information is required? The information required to perform a thorough assessment during any audit is the same regardless of the reason(s) an audit was deemed necessary.

To begin an audit, a letter or memo should be written to the operating company stating precisely what is being requested and why. This request should include, but is not limited to:

Time span of audit;

(If charts used) Flow charts, copies of meter tests/calibrations, analyses, volume statements, and any other data necessary to calculate volume; and

(If electronic flow measurement (EFMs) - flow computers used) Original hourly data, edits, characteristic report, events log, meter tests/calibration, volume statements, and any other data necessary to calculate volume.

Contract Review

The next and very important step will be to obtain a copy of the contract and/or tariff to help identify the extent of controls intended to regulate the flow measurement quality. Each company has its standard contract, defined by using guidelines set by AGA, API, and other organized industry groups. Reviewing these contracts will reveal special provisions pertaining to each individual station. Areas that should be included in the evaluation:

Equipment required;

Meter calibration;

Meter test and sampling frequency;

Adjustment (correction) criteria;

Product quality criteria;

Method for quantity calculation and uncertainty allowance percentage;

Certification frequency on testing equipment;

Statute of limitations;

Pricing and billing specifics; and

Definitions of buyer, seller, contract time, average local atmospheric pressure, pressure base, custody point, etc.

Field Measurement Equipment Review

It is vital for all locations, whether in the main offices, field locations, outside companies, or other departments, to use the same data for their jobs. Verifying the following information can eliminate possible areas of differences:

Station number, name, and location;

Meter type and size(s)—tube and orifice plate sizes for orifice, sensor types, etc.;

Ranges of static, differential, and temperature;

Clock rotation for charts;

Upstream or downstream pressure taps;

Units used, absolute or gauge.

More in-depth reviews may include drawings of the meter site that show the length of the meter tube, and location of taps and direction of product flow and/or physical inspection of the internal condition of flow meters and meter tubes. This will also help determine the location of regulators, valves, dehydrators, scrubbers, storage tanks, production treating and processing units, and any other equipment located near the meter that could directly or indirectly affect measurement uncertainty. In addition, inspection of the internal condition of meters and/or meter tubes can also identify fluid conditions that could affect measurement uncertainty. There may also be a third party check meter at the site for measurement comparison.

Data Review and Comparison

After reviewing the contract, securing the information needed, and determining what equipment is being used, various documents can be compared to spot any inconsistencies. The comparisons should include:

Volume statement comparison (if check measurement used);

Analysis components versus factors actually used;

Unusual variances of daily volumes;

Meter test results adjusted if necessary;

Correct procedure used;

Field reports versus volume statements;

Meter descriptions (tube and plate sizes for orifice metering);

Meter changes and calibrations (and orifice plate changes, if applicable); and

Consistency of dates and times.

Auditing Gas Measurement Systems

Several types of gas audits can be run so the extent of the audit must be determined. (The same general concepts can be applied to liquid audits.) It is important to define the "why and what" involved with the audit. Methods broadly include:

Full comprehensive audit: covers an expanded time span, the contract, all meter test (calibration) reports, volume statements, analysis, gas flow charts (or EFM audit package), and a final report;

Comparison: usually used for sites that have check measurement. Both company's volumes statements are needed, as well as copies of documents showing changes that affect the calculation of volume;

Spot audit: generally used when reviewing discrepancies or known problems; includes the same documents as the comprehensive version, except for a shorter time span and smaller number of items; and

Internal: utilized to ensure that the information used to calculate volumes is correct within the computer system used. Generally, a comparison of field test data versus computer data.

Different steps can be used for internal audit, so the client for which the audit is being conducted should define the desired procedures. Regardless of the approach used, the final objective is the same: to validate information received from the field with data used within the computer system for calculation purposes. Typical information can include:

Prepared form completed by the technician upon physically visiting each meter site;

Most recent meter (calibration) test report; and

Computer-generated list of specified details.

Chart Review

Upon receiving charts, each group should be arranged by date and time. Review and censor each chart before integrating. Some items to look for:

Flow patterns—are they consistent, any downtime?

Irregularities caused by freezes, inspections, testing, ink problems, or missing flow?

Were there notations of fast/slow clocks; high/low zero; recorder problems? And if so, was flow adjusted correctly?

Now charts can be reintegrated. Remember, *do not use ink* as this will obscure the original integration; use either onionskin or dry ink. There are accepted industry standards for integrating charts, but the practices can vary from company to company.

Upon completing reintegration, compare integrator readings and volumes created from the audit against original readings and volume statements. Discrepancies can be caused by differences in integration practices, invalid factors, hours of flow, etc. If the difference is greater that the contractual allowable tolerances, an adjustment may be warranted. If an adjustment is not warranted, the difference should be documented and included in an accounting for "lost and unaccounted."

The major difference between chart review and EFM review is how information is viewed. Charts have a visual flow pattern. EFMs provide numerical patterns. But the concept is the same. Look for abnormalities in the flow pattern, factors used for calculations, characteristics, event logs, and editor change log. When an audit cannot be completed with the above information, it may be desirable to request the raw, original unedited data.

Most companies incorporate a recalculation program that enables them to recreate the daily volumes that can be used for comparison purposes. Again, the contractual agreement should address the tolerance for allowed discrepancies.

Auditing Liquid Measurement

Auditing liquid measurement requires combining a solid knowledge of the measurement process with sound auditing techniques. Just as with the gas audit, steps in a liquid audit should:

Determine the objective;

Identify design procedures;

Collect appropriate evidence;

Perform tests necessary to confirm data accuracy;

Form a conclusion; and

Institute corrective action for problems found.

A liquid audit may involve tank gauging, truck loading systems, and/or barge loading systems. It will obviously relate to measurement systems designed specifically for liquid measurement. The most common meter types encompassed are orifice, liquid turbine, and PD, with growing use of ultrasonic and Coriolis meters—and much less frequently, vortex-shedding.

Finalizing the Audit

If a complete gas or liquids system audit review still leaves doubts, ask questions. Review documents again. Do whatever is required to satisfy the unanswered questions. If adjustments are warranted, a letter should be written to the operating company with specific explanations and findings plus an offer of supporting documents. It is important for the auditor to be open-minded and fair; it should not matter who is at fault if a mistake is found. The important thing is to find and correct the error.

Conclusion

Because we are human and susceptible to mistakes, ensuring accuracy is the primary focus of any audit. It is important to verify that contractual agreements are being met and thus properly account for product measured through any system.

Custody transfer measurement begins with a contract between two parties that specifies the data needed to choose a metering system. To get the most accurate measurement required to minimize settlement problems, the maintenance and operation of the system must be controlled so that the potential performance capabilities of the meter may be realized in service. An audit is a controlled study to demonstrate these performance capabilities.

A properly run audit will protect the client's interest and revenues. These contractual requirements also form the basis for resolving any disputes or litigation between seller and buyer. Properly conducted, audits should improve business relations by imparting useful knowledge to all concerned.

INDEX

Note: Page numbers followed by "*f*", "*t*" and "*b*" refers to figures, tables and boxes respectively.

Edwards Brothers Malloy
Ann Arbor MI. USA
June 13, 2014